Process Intensification 2016

Topical Conference at the 2016 AIChE Annual Meeting

San Francisco, California, USA
13-18 November 2016

ISBN: 978-1-5108-3423-1

Printed from e-media with permission by:

Curran Associates, Inc.
57 Morehouse Lane
Red Hook, NY 12571

Some format issues inherent in the e-media version may also appear in this print version.

Printed by Curran Associates, Inc. (2017)

For permission requests, please contact AIChE
at the address below.

AIChE
120 Wall Street, FL 23
New York, NY 10005-4020

Phone: (800) 242-4363
Fax: (203) 775-5177

www.aiche.org

Additional copies of this publication are available from:

Curran Associates, Inc.
57 Morehouse Lane
Red Hook, NY 12571 USA
Phone: 845-758-0400
Fax: 845-758-2633
Email: curran@proceedings.com
Web: www.proceedings.com

TABLE OF CONTENTS

464945 A Novel and Systematic Method for Process Intensification

Abstract:

Process intensification combines multiple operations in a single unit, and leads to substantially smaller, cleaner, safer, and more energy-efficient technologies [1]. It has been shown that intensification represents a limit case of tight integration through significant material recycling [2]. Although there are established methods for process synthesis and integration, few methods exist for the systematic identification and selection of intensification opportunities. Current superstructure-based process synthesis relies on pre-specified equipment configurations and is unable to automatically construct novel equipment configurations for intensification. Phenomena based process intensification [3-5] allows a bottom-up approach for the identification, screening and incorporation of intensification opportunities, but this approach often requires sequential or decomposition based solution strategies to solve large models, which can lead to suboptimal solutions.

In this work, we propose a unified process synthesis and intensification approach that leverages on an original representation of the flowsheet superstructure. Unlike a classical flowsheet superstructure which encompasses all feasible sequences of unit operations (e.g., reactors, separators, movers, etc.), our proposed superstructure is an ensemble of building blocks, where each block represents a unit use of a material and nexus between the blocks are achieved by intra-block mass and energy transfer. An intensified unit is realized by selecting and assembling multiple neighboring blocks with different functionalities. While reaction, mixing, heating and cooling are achieved within the block, separation related phenomena (e.g., phase separation, phase-contact, etc.,) are achieved via the boundaries between two neighboring blocks. This enables to include many intensification alternatives, including divided wall column distillation, reactive distillation, reactive absorption and membrane reactor, within the same superstructure without any *a priori* postulation.

The overall process intensification model is formulated as a single mixed-integer nonlinear optimization (MINLP)

problem and is solved to optimality. Through a range of applications, we demonstrate that the proposed intensification method is capable of generating innovative and intensified configurations and obtaining optimal process conditions for given feed and product specifications.

References:

[1] Reay, D.; Ramshaw, C.; Harvey, A. Process intensification: engineering for efficiency, sustainability and flexibility. Waltham, MA: Butterworth-Heinemann, 2013.

[2] Baldea, M. From process integration to process intensification. *Computers & Chemical Engineering*2015, 81:104â€"114.

[3] Arizmendi-Sanchez, J. A., and Sharratt, P. Phenomena-based modularisation of chemical process models to approach intensive options. *Chemical Engineering Journal* 2008, 135(1):83-94.

[4] Lutze P, Babi DK, Woodley JM, Gani R. Phenomena based methodology for process synthesis incorporating process intensification. *Industrial & Engineering Chemistry Research* 2013, 52(22):7127â€"44.

[5] Babi, D. K., Holtbruegge, J., Lutze, P., Gorak, A., Woodley, J. M., and Gani, R. Sustainable process synthesisâ€"intensification. *Computers & Chemical Engineering* 2015, doi:10.1016/j. compchemeng.2015.04.030.

469723 High-Resolution Process Design: Flowsheet Optimization with Embedded High Fidelity Unit Models

Abstract:

High-resolution process design: Flowsheet optimization with embedded high fidelity unit models

R.C. Pattison and M. Baldea

McKetta Department of Chemical Engineering, The University of Texas at Austin, Austin, TX 78712

Optimal process design is key to maximizing the profitability of chemical plants. Due to the large upfront capital investments required to construct new processes or retrofit existing plants, it is imperative that the design is optimized using modern gradient-based optimization tools coupled with detailed models of the processing units present in the flowsheet.

Equation-based simulation and optimization have witnessed substantial progress, and most current simulators offer equation-oriented simulation options as well. In this presentation, we illustrate a novel equation-oriented simulation tool which allows for high-fidelity models of select unit operations (including reactors, multi-stream heat exchangers) to be easily embedded into the process flowsheet model. These models capture phenomena occuring over multiple length scales, including smaller dimensions (e.g., diffusion and reaction in a catalyst pellet) which are typically not considered in flowsheet simulation or optimal design calculations.

Our framework is based on a pseudo-transient representation of the unit operation models developed in our previous work [1-3]. The pseudo-transient modeling framework relies on a systematic reformulation of the material and energy balances into a differential and algebraic equation model having the same steady state solution to that of the original algebraic equation model. The (typically) partial differential equations associated with detailed unit models lend themselves naturally to a pseudo-transient reformulation which can then be incorporated into the process flowsheet model. This, in turn, enables the detailed steady state unit designs to be optimized simultaneously with the design of the entire process flowsheet. To this end, we use a time relaxation-based optimization algorithm previously developed in our group [1].

The framework will be illustrated with industrial case studies, including the design optimization of an intensified dimethyl ether production process featuring a dividing wall distillation column and the optimal design of a natural gas liquefaction process.

References

[1] Pattison, R.P.; Baldea, M. Equation-oriented flowsheet simulation and optimization using pseudo-transient models. *AIChE J.*, **2014**, 60, 4104-4123.

[2] Pattison, R.P.; Baldea, M. Multistream heat exchangers: Equation-oriented modeling and flowsheet optimization. *AIChE J.*, **2015**, 61, 1856-1866.

[3] Pattison, R.P.; Gupta, A.M.; Baldea, M. Equation-oriented optimization of process flowsheets with dividing-wall columns. *AIChE J.*, **2016**, 62, 704-716.

456412 Process Intensification of Multicomponent Distillation Configurations Using Minimum Additional Number of Heat and Mass Integration Sections

Abstract:

We present a powerful and simple-to-use fact about heat and mass integration to consolidate distillation columns in a multicomponent distillation configuration. First, thermal couplings are introduced to all intermediate transfer streams in the distillation column. And then, distillation columns associated with a lighter final product reboiler and a heavier final product condenser are consolidated to produce a heat and mass integrated configuration. We call this novel set of configurations the HMP configurations.

Coupled with the already developed techniques[1] that enumerate all basic and thermally coupled distillation configurations, a systematic way of synthesizing all HMP configurations for *n*-component separations is introduced here. Also, we compare the energy savings of HMP configurations with the well-known fully thermally coupled (FTC) configuration using the state-of-the-art Global Minimization Algorithm (GMA)[2] developed based on Underwood equations. GMA provides a global optimization based rank-list of basic and thermally coupled configurations according to minimum total vapor duty requirement under minimum reflux conditions for separating any ideal or near-ideal multicomponent mixture. We demonstrate through case studies that HMP configurations can have very similar and sometimes even the same minimum total vapor duty requirement as the FTC configuration, while using far less number of column sections, intermediate transfer streams, and thermal couplings than the FTC configuration.

These findings not only intensify the search space of distillation configurations, but also provide industrial practitioners with insights in designing energy efficient distillation configurations that are easy to operate and control.

[1] Shah VH, Agrawal R. A matrix method for multicomponent distillation sequences. AIChE J. 2010;56(7):1759â€"1775.

[2] Nallasivam U, Shah VH, Shenvi AA, Huff J, Tawarmalani M, Agrawal R. Global optimization of multicomponent distillation configurations: 2. Enumeration based global optimization algorithm. AIChE J. 2016;62(6):2071â€"2086.

469729 Novel Optimization-Based Approach for Process Design and Intensification of High-Dimensional Modular Systems

Abstract:

In the recent years, due to the advent of the shale gas revolution, the US has incremented the annual production of processed natural gas by approximately 21% from 16.3 to 19.8 trillion cubic feet[1]. To accelerate the potential benefits brought by the abundant amounts of shale gas in the country, approaches are needed for the optimization and techno-economic viability analysis of novel natural gas utilization processes that produce heat, power, fuels and chemicals. However, the design of such processes is a challenging task as these systems are typically operated in a highly constrained and integrated environment that is represented by complex large-scale and nonlinear models. In this presentation, a novel optimization-based operability approach for process design and intensification of high-dimensional and modular energy systems is proposed. These modular units if enabled have the potential to transform the US economy by utilizing local stranded gas without the need of building expensive pipelines.

Process operability was originally developed as an approach for the design and control of complex chemical processes[2]. This approach has been applied to processes that are described by high-dimensional linear models[3,4]. For nonlinear systems, an operability-based approach for design and intensification of low-dimensional membrane reactor systems has also been proposed[5,6]. However, operability methods for nonlinear systems are currently limited by the problem size that they can address. The proposed approach bridges this gap in the literature by addressing the challenges of process nonlinearity and model size. The proposed method also broadens the scope of the traditional path of the operability approaches for design and control, mainly oriented to obtain the achievable output set (AOS) from the available input set (AIS), and compare the computed AOS to a desired output set (DOS). In particular, an optimization algorithm based on nonlinear programming tools is formulated for the high-dimensional calculations of the desired input set (DIS) that is feasible considering process constraints, performance levels and intensification targets.

To illustrate the effectiveness of the developed method, two modular systems are addressed: i) a membrane reactor (MR) for the direct methane aromatization (DMA) conversion to benzene and hydrogen; and ii) a natural gas combined cycle (NGCC) system for heat and power generation. Results on the application of this novel method as a tool for process intensification show reduction of the DMA-MR footprint (â‰ˆ77% reactor volume and 80% membrane area reduction) for an equivalent level of performance, when compared to the base case. These results indicate that the novel approach can be a powerful tool for enabling the realization of the concept of the modular economy.

References

1. EIA (2016). U.S. Energy Information Administration. Natural Gas Report. Available at: http://www.eia.gov/dnav/ng/ng_sum_lsum_dcu_nus_m.htm
2. Vinson D. R., Georgakis C. A new measure of process output controllability. J. Proc. Cont. 2000, 10(2-3), 185-194.
3. Lima F. V., Georgakis C. Design of output constraints for model-based non-square controllers using interval operability. J. Proc. Cont. 2008, 18(6):610-620.
4. Lima F. V., Georgakis C. Input-output operability of control systems: the steady-state case. J. Proc. Cont. 2010, 20(6):769-776.
5. Carrasco J. C. and Lima F. V. Nonlinear operability of a membrane reactor for direct methane aromatization. In Proceedings of the 2015 IFAC ADCHEM Symposium, Whistler, Canada, June 2015.
6. Carrasco J. C. and Lima F. V. Novel operability-based approach for process design and intensification: application to a membrane reactor for DMA. Submitted for publication, 2016.

470719 Towards Optimal and Sustainable Operation of Separation Processes: The Computational Approach

Abstract:

Separation processes (such as Simulating Moving Bed, Pressure Swing Adsorption Systems and chromatographic processes) are widely applied in the chemical/biochemical industry for the separation and/or purification of various components [1-3]. Commonly, chromatographic separations are assessed by their product recovery rate, as well as the purity of the end product, with the latter being of vital importance. Consequently it is essential to guarantee that the process runs under optimal operating conditions, achieving high recovery rates of products that meet the desired specifications. However, downstream processing is coupled with high capital and operational costs that are significantly increased as a function of the feed stream produced by the upstream [4]. Current industrial needs suggest the development of sustainable processes, characterized by lower investment and/or operation costs. To answer to global demand and competition arising from novel solutions such as single-use technologies [5, 6], separation processes and particularly preparative chromatography, requires significant advancements.

Along this line, in this work we focus on the development of advanced control strategies for a semi-continuous, chromatographic separation process [7] aiming to increase both product purity and yield. In this framework we examine a highly nonlinear, periodic system and we design novel control strategies to ensure that optimal operation is maintained. The controllers are designed and tested following our recently presented PAROC framework that comprises four steps: (i) development of a high-fidelity process model, (ii) approximation of the complex, process model, (iii) design of a multi-parametric controller and (iv) â€˜closed-loopâ€™, in-silico validation of the controller against the process model. The controllers are tested in-silico against the high-fidelity process model and demonstrate an overall satisfactory performance, tracking the predefined setpoints and providing input profiles of reduced utility consumption. In addition, the proposed control scheme accounts for variable feeding composition, thus enabling the downstream processing to operate independently from the mixture composition resulting from the upstream.

Acknowledgements

The authors would like to thank Mr R. Oberdieck and Miss A. Quiroga-Campano for their contribution in the assessment of the controller validation. The authors would also like to thank Mr. Fabian Steinebach & Prof. M. Morbidelli from ETH Zurich, as well as Dr. Thomas Mueller-Spaeth from ChomaCon AG for their valuable input on the understanding of the MCSGP process. Financial support from the European Commission (OPTICO/G.A. No.280813) & Texas A&M University are also gratefully acknowledged.

References

1. Subramani, H.J., K. Hidajat, and A.K. Ray, *Optimization of Simulated Moving Bed and Varicol Processes for Glucoseâ€"Fructose Separation.* Chemical Engineering Research and Design, 2003. **81**(5): p. 549-567.

2. Khajuria, H. and E.N. Pistikopoulos, *Dynamic modeling and explicit/multi-parametric MPC control of pressure swing adsorption systems.* Journal of Process Control, 2011. **21**(1): p. 151-163.

3. Müller-Späth, T., et al., *Chromatographic separation of three monoclonal antibody variants using multicolumn countercurrent solvent gradient purification (MCSGP).* Biotechnology and Bioengineering, 2008. **100**(6): p. 1166-1177.

4. Gronemeyer, P., R. Ditz, and J. Strube, *Trends in Upstream and Downstream Process Development for Antibody Manufacturing.* Bioengineering, 2014. **1**(4): p. 188-212.

5. Shukla, A.A. and U. Gottschalk, *Single-use disposable technologies for biopharmaceutical manufacturing.* Trends in Biotechnology, 2013. **31**(3): p. 147-154.

6. Xenopoulos, A., *A new, integrated, continuous purification process template for monoclonal antibodies: Process modeling and cost of goods studies.* J Biotechnol, 2015. **213**(0): p. 42-53.

7. Krättli, M., F. Steinebach, and M. Morbidelli, *Online control of the twin-column countercurrent solvent gradient process for biochromatography.* Journal of Chromatography A, 2013. **1293**: p. 51-59.

454543 Multi-Scale Computer Aided Synthesis–Design–Intensification Method for Sustainable Hybrid Solutions

Abstract:

Process synthesis-design refers to identification of the processing route to reach a desired product from a specified set of raw materials, design of the operations involved in the processing route along with calculations of utility requirements, waste generations and emissions to the environment. Process intensification refers to strategies by which new processes can be designed or existing processes can be redesigned to be more efficient, compact and sustainable.

Therefore, in order to achieve sustainable process design it would be logical to perform process synthesis-design and process intensification together.

In this work, a four stage hierarchical decomposition solution strategy is proposed to generate sustainable hybrid solutions. The framework operates at different scales, the unit operations scale, the task scale and the phenomena scale. In stage1, process synthesis is performed (at the unit operations scale) using a newly developed computer-aided flowsheet design method (CAFD) (Tula et al, 2015), which applies principles of computer-aided molecular design to the synthesis and design of process flowsheets. That is, use process-groups representing different unit operations (reactor, distillation, flash, crystallization, etc.), bonds representing streams and/or recycles, rules for chemical feasibility also representing process flowsheet feasibility and sum of group contributions representing the performance of the flowsheet that can later be used to quickly screen the alternatives and to generate the optimal processing route. In stage 2, the base case design is first established based on the generated processing route. This base case design is further analyzed to identify process hotspots (process bottlenecks) using a comprehensive indicator based analysis incorporating process economics, sustainability measures (Carvalho et al., 2013) and life cycle assessment factors (Kalakul et al., 2014). These indicators corresponding to the hotspots are translated to design targets that are targeted in the next stage to achieve more sustainable and intensified designs. In stage 3,

integrated task phenomena based synthesis-intensification method is applied (Babi et al., 2015). The main idea here is to generate all the phenomena involved in the process by breaking the tasks involved into corresponding phenomena. Later these phenomena scale units are combined using combination rules in order to generate new and/or existing unit operations that constitute the (more sustainable) flowsheet alternatives which satisfy, the process design targets. In this way, truly predictive and innovative hybrid solutions are generated much in the same way as atoms are combined to form molecules with desired properties (that is, analogous to computer-aided molecular design) which are otherwise could not be found from the higher scales. In the final stage, detailed model based simulations and/or experiments are performed to verify and compare the new solutions.

In this presentation, a multi-scale computer-aided framework will be presented together with the corresponding databases, computer-aided methods and tools needed to implement and apply the synthesis-design-intensification method. The application of this framework will be highlighted through case studies generating hybrid solutions for energy efficient processing schemes involving downstream separations and intensified reaction-separations.

References:

[1] Babi, D. K., Holtbruegge, J., Lutze, P., Gorak, A., Woodley, J. M., & Gani, R. 2015. Computers & Chemical Engineering.

[2] Carvalho, A., Matos, H. A., & Gani, R. 2013. Computers & Chemical Engineering, 50, 8–27

[3] Kalakul, S., Malakul, P., Siemanond, K., & Gani, R. 2014. Journal of Cleaner Production

[4] Tula, A. K., Eden M. R., Gani, R. 2015. Computers & Chemical Engineering.

458409 Microwave Assisted Crystallization: Intensification of Direct Nucleation Control

Abstract:

Direct nucleation control (DNC) is a feedback control strategy used to control the size distribution of crystals (one of the many possible control objectives) during a crystallization process. DNC has been shown for maintaining a narrow size distribution by removal of unwanted fine particles. Application of DNC for obtaining desired size distribution is based on alternating dissolution and growth cycles, achieved via temperature cycling and triggered as a result of deviations in the particle count set point (measured by in-situ measurement PAT tool).

However, a major draw-back of the DNC implemented in conventionally heated/cooled reactors is that the temperature cycles needed to dissolve the nuclei considerably increases the process/batch time. However recently it has been demonstrated that by application of microwave fields for heating, the same control performance can be achieved with a 50% reduction in process/batch time. The study showing the intensification of the DNC in terms of process time by integrating microwave fields into a crystallizer was performed in a 1 litre jacket vessel in which the slurry was circulated via an external loop through a microwave cavity. However a better application would be when microwave fields are truly introduced inside the crystallizer thus eliminating external loops for slurry circulation which can be problematic.

In this study we present DNC experiments using a microwave integrated crystallizer. The crystallizer utilizes internal transmission line technology made available by Sairem SAS which enables direct introduction of microwave fields into the slurry. In this study we demonstrate cooling crystallization with DNC control and address key challenges in process integration and implementation, namely: Evaluation and optimization of the microwave heating to ensure homogeneous temperature distribution in the system, Optimization of the reactor configuration for safe introduction of PAT tools essential for DNC feedback (e.g FBRM), by avoiding the exposure to high intensity MW field and

evaluation of the performance of microwave assisted DNC in a cooling crystallization process.

467819 A Systems Approach for the Development of Osn Membrane Cascades

Abstract:

The processes involved in separating the components of chemical mixtures into pure or purer forms, such as distillation, account for 10-15% of the worldâ€™s energy consumption. Methods to purify chemicals that are more energy efficient could, if applied to the U.S. petroleum, chemical and paper manufacturing sectors alone, save 100 million tonnes of CO_2 emissions and US$4 billion in energy costs annually [1]. Membrane-based separation methods, or other non-thermal ones, can be an order of magnitude more energy efficient than heat-driven separations that use distillation, but these methods are underdeveloped [2].

Given the recent advances in organic solvent nanofiltration (OSN) membranes [3], the focus of this presentation is on OSN membrane cascades for separating organic/hydrocarbon mixtures. As a case study, we investigate the design of a lab-scale membrane cascade for purifying a binary mixture of heptane and hexadecane, with a 75-25 weight fraction. The main objective is to maximize heptane purity in the permeate stream by using a maximum of 10 identical membrane units, each having an effective area of 14 cm.

In a first step, we use a superstructure optimization approach based on mixed-integer nonlinear programming (MINLP) to determine the most promising cascade configurations, in terms of the number of stages in the cascade, the number of membrane units in each stage, and the interconnections thereof, along with certain operating parameters [4,5]. Constraints are imposed for limiting the transmembrane pressure drop and the stage-cut in each unit. The permeate and retentate flux and composition for each membrane unit are predicted by a classical solution-diffusion model, where the permeability coefficients for heptane and hexadecane are estimated based on dedicated experiments for the membranes at hand, and the activity coefficients of heptane and hexadecane are approximated using the UNIFAC method for different mixture compositions. An optimal 3-stage cascade obtained with this approach is shown on the figure below, with GAMS 24.5.6 and the global optimizer BARON 15.9 used to solve the MINLP problem.

In a second step, we implement the optimised OSN membrane cascades experimentally in order to verify the predictions. For the 3-stage cascade shown on the figure, the mismatch between the model predictions and the measurements remains small, within a few percents, thereby validating the model-based optimization approach.

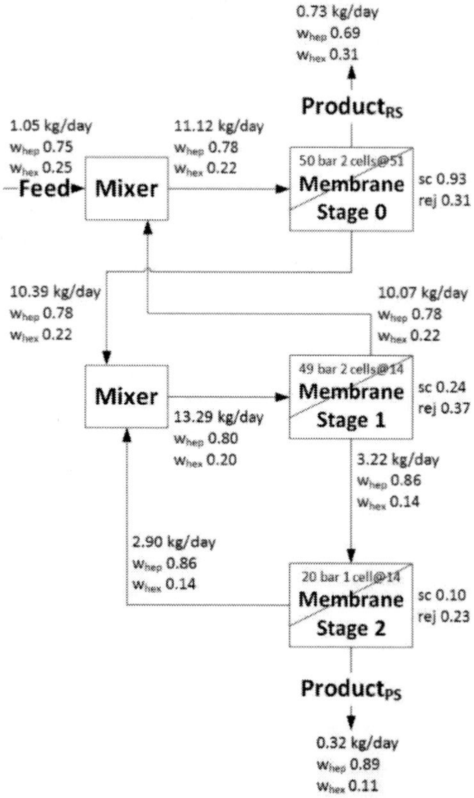

References:

1. U.S. Department of Energy, Advanced Manufacturing Office, "Bandwidth Study on Energy Use and Potential Energy Saving Opportunities in U.S. Petroleum Refining", June 2015.

2. D.S. Sholl, R.P. Lively, "Seven chemical separations to change the world," Nature 532:435-43, 28 April 2016. U.S. Petroleum Refining (US Dept. Energy, 2015).

3. P. Marchetti, M.F.J. Solomon, G. Szekely, A.G. Livingston, "Molecular separation with organic solvent nanofiltration: A critical review," Chemical Reviews 114:10735-10806, 2014.

4. R.H. Qi, M.A. Henson, "Membrane system design for multicomponent gas mixtures via mixed-integer nonlinear programming," Computers & Chemical Engineering 24:2719-2737, 2000.

5. V.S.K. Adi, M. Cook, L.G. Peeva, A.G. Livingston, B. Chachuat, "Optimization of OSN membrane cascades for separating organic mixtures," Proc. 26th European Symposium on Computer Aided Process Engineering (ESCAPE), 12-15 June 2016, PortoroÅ¾, Slovenia.

475310 Introduction to Altsep: Sustainable Separation Processes Roadmap Review

Abstract:

Creating a sustainable future requires sustainable manufacturing processes. This is especially true for the separation processes central to chemical manufacturing. Conventional fluid separations account for over a third of energy use and over half of the capital investment of chemical plants. To pave the way toward transforming the energy efficiency of fluid separations, the ACS Green Chemistry Institute (GCI) has partnered with AIChE to coordinate the development of a NIST-sponsored innovation roadmap. Based on a series of workshops, the draft AltSep roadmap has identified a number of research, development, and demonstration initiatives. These initiatives are critical for establishing a molecular property-based system for the industrial selection, simulation, and design of energy-efficient separation processes and accelerating the industrial application of such alternative separation processes.

A series of speakers will present on the following topics:

- Background and need for the roadmap,
- System to synthesize and design alternative separation processes, and
- Challenges to overcome barriers to adoption of sustainable separation solutions.

Following these presentations, feedback will be solicited from the audience to improve the quality of the roadmap findings.

451921 Reaction Intensification in Coal Tar Refining

Abstract:

Reaction intensification in coal tar refining

SHASHATY A.[1], GUILLOU L.[2]

1. McGill University, 845 Rue Sherbrooke O, QC H3A 0G4, Montréal, Canada

2. Fives Proabd, 20C rue de Chemnitz – 68100 Mulhouse, France

Processing of coal tar oils has been a way of choice for the recovery of dozens of cyclic, aromatic or polyaromatic compounds: phenol, cresols, indane, coumarone, pyridine bases, thiophenols, carbazole, [1]… The associated coal tar distillation industry has been the flag ship for the development of industrial chemistry up to the 50's. If this industrial area is nowadays considered as being mature the well established associated technology portfolio [2] is ready to evolve. Some opportunities offered by process intensification give access to new options for the tar refiners. Beyond the most obvious opportunities for decreasing the refinery footprints by using some smaller and less energy demanding equipments that are offered by intensified devices, the implementation of these new technologies offers new production alternatives for recovery of chemicals of interest which were given up due to uncompetitive operational costs vs. synthetic routes.

A typical example is considering the so called tar acids. This family of compounds gathers phenol and its homologue molecules (cresols and xylenols). The traditional recovery route is based on a series of reactions with strong bases and strong acids. In a first step, the oil is treated with caustic soda to recover an aqueous phenolate liquor which is then reacidified to reform an organic phenol blend. Such a process is usually either a batch process or a continuous Lurgi phenoraffin process. In the first case, due to the nature of the chemicals involved, a significant investment is required: glass lined reactors, graphite heat exchangers… Serious industrial safety concerns are also raised. This leads to the definition of an economical bottom line which restraint these applications to large coal tar refineries (beyond 400 ktpy whereas most refineries are nowadays within 200-300 ktpy size range). As a matter of fact, shall the phenolate liquor be removed for some process constraints in the distillation train, they are either qualified as a waste or as a low value byproduct which shall be sold to another industry. In such a latter case, these streams are collected throughout many refineries and treated in a single central facility; thanks to economies of scale, these downstream treatment centers can afford moving to continuous processes such as phenoraffin.

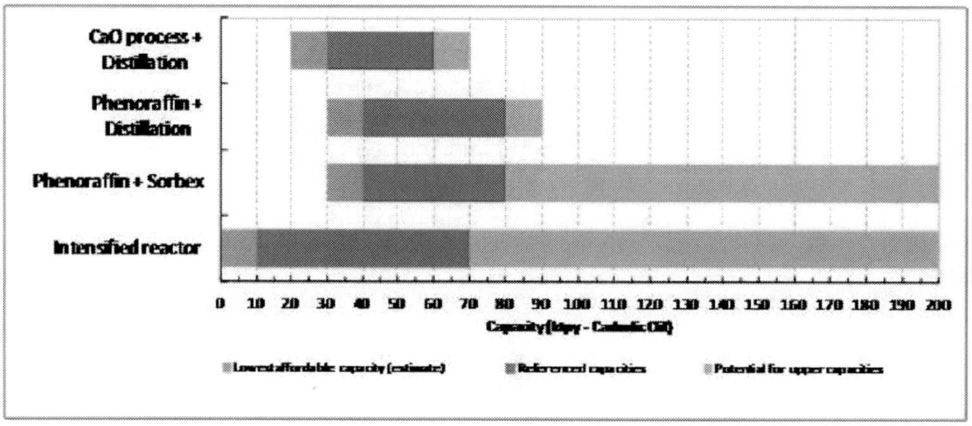

The potential for the recovery of phenols and other tar acids has been investigated at laboratory scale in a microchannel reactor. The process consisted in a two steps:

- A neutralization of a carbolic oil/naphtha fraction with diluted caustic soda
- An acidification with sulfuric acid for forming back phenols for sodium phenolates.

These reactions show some distinctive features such as:

- Large exothermicity both due to the reaction and the dilution of acid or base in the media
- Extremely fast kinetics
- Undesirable formation of bright red coloring agent for uncontrolled temperature rise
- Polyphasic systems

The use of microchannel reactor has been found to be a good alternative to the traditional processing routes. High control of the exothermy has been achieved and phenol recovery has allowed achieving satisfactory yields with decreased water consumption. This technology opens a route for compact reactor systems perfectly compliant with continuous operation in refineries even for small capacities. This new generation of compact reactors opens opportunities for renewal of coal tar refining business model, assisting refiners into a move toward higher value chemicals whatever the size of the distillation train might be.

[1] Franck H.-G., National meeting of American Chemical Society, 13 Jan. 1963, Vol 7:1 Conference 143.

[2] Franck H.-G., Stadelhofer J.W., in "Industrial aromatic chemistry", 1988, Springer-Verlag, Berlin-Heidelberg-N.-Y.

Keywords: reaction intensification, compact reactor, coal tar.

456936 Application of Dynamic Heat Exchanger Design in Natural Gas Liquefaction Process

Abstract:

The paper discusses transient Aspen HYSYS modeling and optimization of two natural gas liquefaction processes and identifies the rate-limiting components during load variations. The optimized model for both processes provides details for comparison. Novel flowrate variations included in this investigation drive transients responses of all units, especially compressors and heat exchanger. Heat exchangers commonly represent the most sensitive components to transients. This sensitivity decreases when using new heat exchanger designs and control methods. Model-predictive control (MPC) responses compare favorably with results using traditional controls. Transient efficiency graphs for both designs illustrate improvements during model predictive control. These new controls and designs optimize natural gas (NG) consumption and liquefied natural gas (LNG) production.

451752 Implementation of Process Intensification (PI) in the Removal of Carbonyl Compounds from the Water/Ethanol Solvent Mixture

Abstract:

Carbonyl compounds are frequently observed ingredients in biochemical, food, chemical and pharmaceutical industries. The high reactivity of such compounds is a great advantage if aldehydes/ketones are important contributors in the synthetic routes towards the final products. However, if the carbonyl compounds act as impurities, they must be removed from the reaction mixture, due to the high tendency of forming genotoxic impurities (GTIs).

Removal of carbonyl compounds could be performed in several ways. Applications of adsorption techniques by using alumina, titanium, or activated carbon, are well-known and even commercially applicable in the petrochemical industry. Implementation of in- or ex-situ removal with pervaporation has also achieved significant advantages with the main application in fermentation processes. Furthermore, combinations of derivatization and separation processes are known, such as reactive extraction with amines or even solid-phase extraction with 2,4-dinitrophenylhidrazine (DNPH). Some new techniques are additionally getting popular, such as applications of extractant impregnated resins (EIRs).

The main aim of this work is to remove benzaldehyde from a solvent mixture consisting of 70 % water and 30 % of ethanol (volumetric ratio). Such a solvent mixture indicates formation of acetals, hemiacetals and hydrates with carbonyl compounds, and could also promote electrophilic and nucleophilic substitution reactions of aldehydes/ketones with other compounds in the reaction mixture. The main approach is focused on finding suitable derivatization agents that would convert carbonyl compounds into new impurities that are harmless. The screening procedure was therefore performed initially with primary/secondary/tertiary amines, inorganic salts, organic acids and their anhydrides. The most desired option was found to be tris(hydroxymethyl)aminomethane (TRIS), which converts carbonyl compounds into oxazolidines. A Process Systems Engineering (PSE) approach including property prediction simulations indicated that such products are more inert compared to carbonyl compounds and additionally more suitable for removal by applying other separation processes.

TRIS was used in a process development study in order to remove ppm levels of benzaldehyde. The initial process design was performed in batch mode with off-line real time process monitoring. A Process Intensification (PI) approach was implemented and the movement from batch to continuous processing was achieved by using a laminar tubular reactor. Real-time process monitoring was also moved from off-line with High Performance Liquid Chromatography (HPLC) towards in-line with UV-VIS spectroscopy in order to satisfy requirements defined within the Process Analytical Technology (PAT) guidance. Both types of processes are compared in order to evaluate advantages of implementing PI, PAT and PSE. Further work will include process control studies and automation, as well as considerations of suitable scale up/out possibilities.

Sensitivity Analysis of the Separation Efficiency of a Vacuum Stripping Microchannel Device

Chafika Adiche, Technische Universität Darmstadt, Thermal Process Engineering, Darmstadt, Germany

Abstract

In this work a two dimensional model describing the transport properties in a vacuum stripping microchannel device was developed. Based on this model, a sensitivity analysis of the separation efficiency of the device to its relevant design and operating parameters was conducted. It is shown that the developed model allows to elucidating the impact of the investigated design and operating parameters on the controlling resistance to mass transfer in the whole device including the liquid-vapor membrane contactor and the liquid channel and thus constitutes a useful tool towards the design optimization of the device.

1 Introduction

The implementation of gas / vapor – liquid separation operations like absorption, stripping, and distillation at microscale constitutes an important milestone for the synthesis of multistep microchemical processes [1]. However, over the last two decades the development of separation microchannel devices for such operations has proven to be a challenging task as compared to the well-established microreactors, micromixers and micro heat exchangers. Indeed the high surface-to-volume-ratio prevailing at microscale results in (i) a poor control of the required thermodynamic vapor-liquid equilibrium [2] and in (ii) a high flow resistance particularly in liquid channel (s). Both phenomena result in the reduction of the separation driving force and thus in the drop of the separation efficiency.

Recent investigations showed that liquid repellent vapor-liquid contactors with hydrophobic surface properties may be successfully used in distillation [3], membrane distillation [4-5] and stripping [6-8] based separation microchannel devices. As compared to wetting vapor-liquid contactors [7-8], the liquid repellent vapor-liquid contactors, may have important potentials to reduce the energy consumption of such microchannel devices [9] as well as the flow resistance in the liquid channels [10]; contributing thereby to a better stabilization of the vapor-liquid interface and thus to the enhancement of the separation efficiency of the device.

In this work, the separation efficiency of a vacuum stripping microchannel device is investigated using a model based approach.

The device is constituted of two micromachined rectangular straight channels separated by a flat sheet microporous hydrophobic membrane (contact area: 80 mm × 6 mm) [Figure 1].

The membrane serves as liquid-vapor contactor, which repels the liquid phase and constitutes therefore a support for the liquid-vapor interface at the entrance of the pores. The liquid feed (aqueous solution of 5 wt. % acetone) is circulated through one of the channels. The other channel called permeate channel is kept under vacuum. Acetone and water evaporate and diffuse through the membrane pores. The separation driving force in the device is created by the partial vapor pressure gradient for each component (acetone, water) across the membrane caused by the vacuum in the permeate channel. The permeated vapor phase is then stripped by vacuum outside the permeate channel.

To avoid liquid intrusion into the membrane pores, the pressure difference across the membrane must be maintained below the liquid breakthrough pressure of the membrane [3-5, 11]. Furthermore, the pressure in the permeate channel must be lower than the vapor pressure of the feed [11].

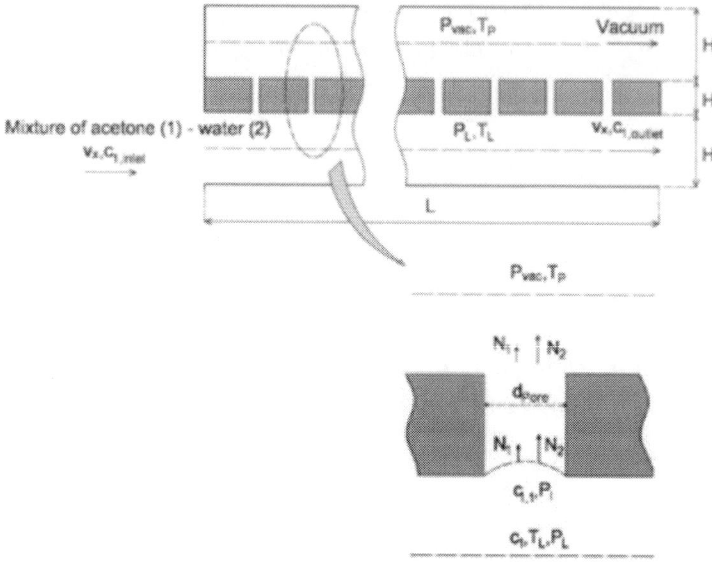

Figure 1. Schematic view of the model of transport properties in a vacuum stripping microchannel device

For the above presented separation device, the following relevant design and operating parameters were identified: (i) the liquid channel depth H_{ch}, (ii) the geometric properties of the considered membrane contactor including porosity ε, mean pore size d_{pore} and thickness H_m, (iii) the feed flow rate or the residence time in the liquid channel τ_{hydr}, (iv) the temperature in the liquid channel T_L and (v) the vacuum pressure P_{vac}.
In this work, two commercial flat sheet microporous hydrophobic membranes provided by Pall (USA) were considered: polyethersulfone polymer cast on a non-woven polyester-support membrane (Supor®-450R) and acryl copolymer membrane on polyester-support (Versapor®-200TR). The contact angles measured for the investigated liquid mixtures on both membrane surfaces were found to be larger than 120 °.

2 Modelling and simulation

Theoretical investigations were carried out for the assessment of the influence of the above mentioned design and operating parameters on the separation efficiency of the vacuum stripping microchannel device at isothermal conditions, and thus to elaborating an optimization approach for the design of the device. For this purpose, a two-dimensional mathematical model describing the transport mechanisms in the device including the liquid channel and the membrane was developed. The model is based on the following assumptions:

a. Since the width of the liquid channel is much larger than the investigated liquid channel depths, the side-wall effects are neglected and the problem may be reduced to two dimensions.
b. For all investigated liquid flow rates, the Reynolds number Re is smaller than 10, so that a hydrodynamically developed laminar flow may be assumed.

c. For all investigated values of vacuum pressure, the Knudson number kn was found to be in the range between 2 and 26, so that the Knudson diffusion may be assumed as the prevailing mass transport mechanism in the membrane pores.
d. Finally, the temperature in the liquid channel is assumed constant and equal to the temperature in the permeate channel i.e. $T_P = T_L$.

Accordingly, the equations of the model are given as follows:

1. The flow velocity profile writes

$$v_x = 1.5\langle v_x \rangle \left(1 - \frac{4y^2}{H_{ch}^2} \right). \tag{1}$$

2. The species conservation equation for acetone is given by
3. $$v_x \frac{\partial c_1}{\partial x} = D^L \left(\frac{\partial^2 c_1}{\partial x^2} + \frac{\partial^2 c_1}{\partial y^2} \right), \tag{2}$$

with the diffusion coefficient D^L is calculated using Wilke-Chang correlation.

4. The diffusion flux for acetone at the liquid-membrane interface is formulated as

$$N_1 = \frac{K_m}{\sqrt{MW_1}} \left(P_{1,1} - P_{vac}\, y_{p,1} \right), \tag{3}$$

where K_m is the permeability coefficient of the membrane [11], $P_{1,1}$ the partial vapor pressure of acetone at the liquid-membrane interface and MW_1 the molecular weight of acetone.
The permeate composition y_p is computed using the double film theory [11].

5. The liquid-vapor equilibrium relationship at the liquid-membrane interface was determined using Wilson model (*Properties Analysis Tool of Aspen Plus*)

$$P_{1,1} = \frac{1.29 c_{1,1}}{c}. \tag{4}$$

6. The boundary conditions for mass transfer are given by

$$c_1\big|_{x=0,y=0} = c_{1,inlet}, \tag{5.a}$$

$$-D^L \frac{\partial c_1}{\partial y}\bigg|_{y=+\frac{H_{ch}}{2}} = N_1, \tag{5.b}$$

$$\frac{\partial c_1}{\partial y}\bigg|_{y=-\frac{H_{ch}}{2}} = 0, \tag{5.c}$$

and $$\frac{\partial c_1}{\partial y}\bigg|_{x=L,\, y=-\frac{H_{ch}}{2};+\frac{H_{ch}}{2}} = 0. \tag{5.d}$$

The previously presented model along with the corresponding boundary conditions has been reformulated using dimensionless variables given by

$$\bar{v}_x = \frac{v_x}{\langle v_x \rangle},\ \bar{y} = \frac{y}{H_{ch}/2},\ \bar{x} = \frac{2x}{H_{ch} Pe},\ \bar{c}_1 = \frac{c_1}{c_{1,inlet}}, \tag{6}$$

with the Peclet number Pe is expressed as

$$Pe = Sc\, Re = \frac{L\, H_{ch}}{2\, D^L\, \tau_{hydr}}. \tag{7}$$

The dimensionless model writes as follows:

- Dimensionless velocity profile:

$$\bar{v}_x = 1.5\left(1-\bar{y}^2\right). \tag{8}$$

- Dimensionless species conservation equation for acetone:

$$\bar{v}_x \frac{\partial \bar{c}_1}{\partial \bar{x}} = \frac{1}{Pe^2}\frac{\partial^2 \bar{c}_1}{\partial \bar{x}^2} + \frac{\partial^2 \bar{c}_1}{\partial \bar{y}^2}. \tag{9}$$

- Dimensionless boundary conditions:

$$\bar{c}_1\Big|_{\bar{x}=0,\bar{y}=0} = 1, \tag{10.a}$$

$$\frac{\partial \bar{c}_1}{\partial \bar{y}}\bigg|_{\bar{y}=+1} = -\frac{H_{ch}\,N_1}{2\,D^L\,c_{1,inlet}}, \tag{10.b}$$

$$\frac{\partial \bar{c}_1}{\partial \bar{y}}\bigg|_{\bar{y}=-1} = 0, \tag{10.c}$$

and $$\frac{\partial \bar{c}_1}{\partial \bar{y}}\bigg|_{\bar{x}=\bar{L},\,\bar{y}=-1:+1} = 0. \tag{10.d}$$

The separation efficiency of the device is measured using the stripping degree. The latter is ranging between 0 and 1, which correspond to 0 and 100 % acetone removal from the liquid phase, respectively.

The degree of stripping σ is determined according to the following equation

$$\sigma = 1 - \frac{\int_{-1}^{+1} \bar{v}_x \bar{c}_1\left(\bar{L}\right) d\bar{y}}{\int_{-1}^{+1} \bar{v}_x d\bar{y}}. \tag{11}$$

The established model with all relevant boundary conditions has been implemented using Finite element Method (Comsol Multiphysics), with mesh of triangular elements. A fine mesh was constructed near the membrane surface where significant concentration gradient is expected. To test the precision provided by the selected mesh-size, a mesh-size independence study was conducted for each investigated *Peclet* number and channel depth.

3 Simulation results and discussion

Based on the elaborated model, a sensitivity analysis of the separation efficiency of the device to the previously mentioned design and operating parameters was conducted.

3.1 Influence of the geometric parameters

In this set of simulations, the channel depth was varied between 50 and 500 μm, while all other relevant design and operating parameters were kept constant.

Figure 2 shows that for the membrane Supor®-450R, decreasing the liquid channel depth until 300 μm affects hardly the device performance such that an almost constant and low stripping degree of about 0.18 was determined. For the same range of liquid channel depths, similar behavior was also obtained using Versapor®-200TR with however a slightly lower stripping degree (0.15). Further decrease of the liquid channel depth gives rise to a greater enhancement of the stripping efficiency for each membrane with however a higher extend for Supor®-450R as compared to Versapor®-200TR. For example for the smallest investigated liquid channel depth of 50 μm, the stripping degree obtained for Supor®-450R was 0.77, whereas that determined for Versapor®-200TR was only 0.67.

These results suggest that for liquid channel depths larger than 300 μm, the molecular diffusion in the feed channel is the rate limiting step for mass transfer in the whole device. However, for a channel depth smaller than 300 μm, the resistance to mass transfer in the membrane becomes more important, so that using the membrane allowing a larger acetone permeate flux, here Supor®-450R (with a larger ratio of the product of membrane pore size and porosity to membrane thickness), leads to a better device performance.
According to these results, Supor®-450R membrane contactor will be considered for the next simulation investigations.

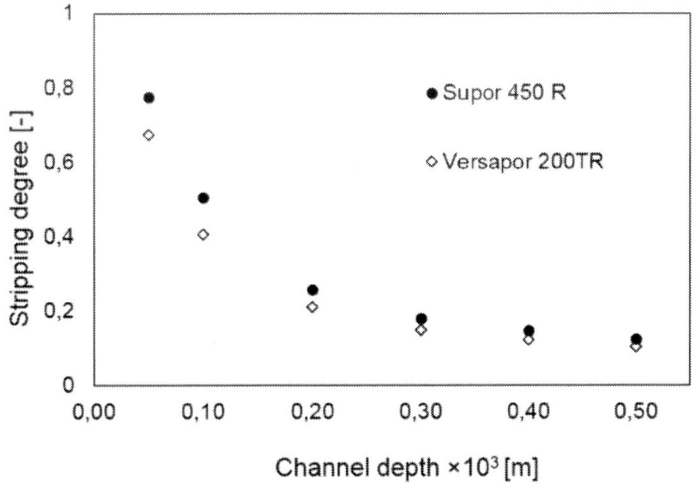

Figure 2. Influence of the channel depth on the stripping degree for different membrane contactors, $x_F = 5$ wt. %, $T_L = 20$ °C, $P_{vac} = 0.01$ bar, $\tau_{hydr} = 5s$.

3.2 Influence of the residence time in the liquid channel

We can see in Figure 3 that for the three investigated channel depths, increasing the residence time in the liquid channel, leads to increasing the stripping degree of the device. Furthermore, the smaller the channel depth, the faster is the increase of the stripping degree until reaching a value of 0.84 for a channel depth of 50 μm and a residence time in the liquid channel of 10 s.

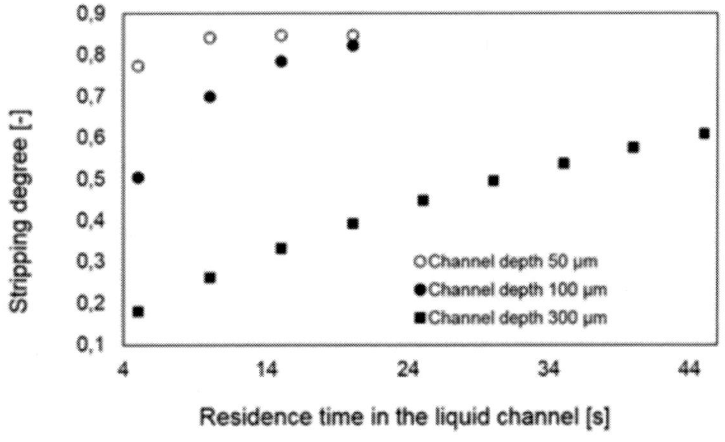

Figure 3. Influence of the residence time in the liquid channel on the stripping degree for different channel depth, $x_F = 5$ wt. %, $T_L = 20$ °C, $P_{vac} = 0.01$ bar, Supor 450R.

This behavior may be attributed to the reduction of the concentration polarization in the liquid channel, resulting therefore in a better mixing of the acetone-water solution (computed diffusion time of acetone in the liquid channel with a depth of 50 μm is about 1.2 s). Consequently, the separation driving force of acetone at the liquid-vapor interface increases, leading therefore to a higher acetone permeate flux, and thus to enhanced stripping efficiency. Similar trend was also reported by Chasani et al. [6] for the stripping of dilute toluene from water using a hydrophobic vapor-liquid contactor into a microchannel device with dry nitrogen as stripping agent.

3.3 Influence of the vacuum pressure

Figure 4 indicates that for each investigated temperature, increasing the vacuum pressure while maintaining all other design and operating parameters constant, results in decreasing the stripping degree. That results in the decrease of the separation driving force, thus in the reduction of the acetone permeate flux and finally in the drop of the stripping efficiency. Furthermore, it is shown that for a given vacuum pressure, increasing the feed temperature improves substantially the stripping efficiency (due mainly to the exponential increase of the vapor pressure of the feed with the temperature), along with broadening the vacuum pressure range.

Similar behavior was also found for the influence of varying the vacuum pressure along with the feed composition, with however the difference that increasing the acetone concentration in the feed for a given vacuum pressure [Figure 5] improves only slightly the stripping efficiency due to the Linear increase of the vapor pressure of the feed with the feed composition.

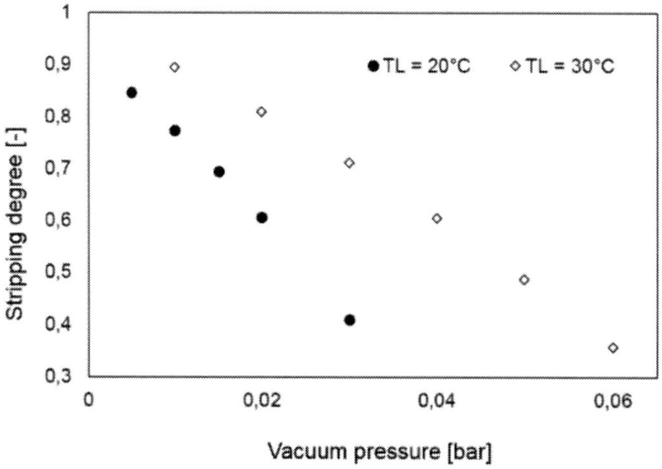

Figure 4. Influence of the vacuum pressure on the stripping degree for different temperatures in the device, $x_F = 5$ wt. %, $\tau_{hydr} = 5s$, $H_{ch} = 50$ μm, Supor 450R.

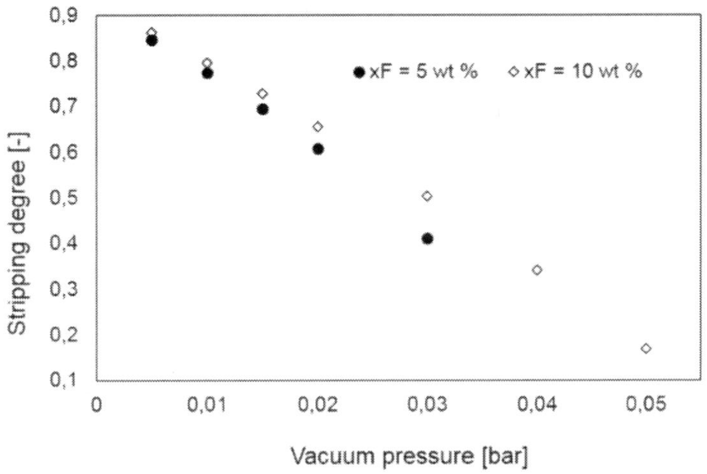

Figure 5. Influence of the vacuum pressure on the stripping degree for different feed compositions, $T_L = 20$ °C, $\tau_{hydr} = 5s$, $H_{ch} = 50$ µm, Supor 450R.

References

1. Pastre, J. C.; Browne, D. L.; Ley, S. V. (2013), '' Flow chemistry syntheses of natural products'', *Chem. Soc. Rev.*, 42, pp. 8849-8869.
2. Hartman, R. L.; Sahoo, H. R.; Yen, B. C.; Jensen, K. F. (2009), ''Distillation in microchemical systems using capillary forces and segmented flow'', *Lab Chip*, 9, pp. 1843-1849.
3. Zhang, Y.; Kato, S.; Anazawa, T. (2009), ''Vacuum membrane distillation on a microfluidic chip'', *Chem Commun.*, pp. 2750-2752.
4. Adiche, C. and Sundmacher, K. (2010), ''Experimental investigation of a membrane distillation based micro-separator'', *Chem. Eng. Proc.: Process Intensification*, 49 (4), pp. 425-434.
5. Adiche, C. and Sundmacher, K. (2010), ''Selection of membrane contactors used in membrane distillation based micro-separator'', *Chem. Eng. Transactions*, 21, pp. 361-366.
6. Chasanis, P.; Kehrmann, K.M.; Kern, J.; Zecirovic, R.; Grünewald, M.; Kenig, E. Y. (2011), '' Investigation of a microstructured high efficiency contactor'', *Chem. Eng. and Process.*, 50, pp. 1244-1251.
7. Constantinou, A.; Ghiotto, F.; Lam, K. F.; Gavriilidis, A. (2014), ''Stripping of acetone from water with microfabricated and membrane gas-liquid contactors'', *Analyst*, 139, pp. 266-272.
8. Cypes, S. H. and Engstrom, J.R. (2004), ''Analysis of a toluene stripping process: a comparison between a microfabricated stripping column and a conventional packed tower'', *Chem. Eng. J.* 101, pp. 49-56.
9. Wootton, R. C. R. and de Mello, A. J. (2004), ''Continuous laminar evaporation: micron-scale distillation'', *Chem. Commun.*, pp. 266-267.
10. Ou, J.; Perot, B.; Rothstein, J. P. (2004), '' Laminar drag reduction in microchannels using ultrahydrophobic surfaces'', *Physics of Fluids*, 16, 12, pp. 4635-4643.
11. Bandini, S.; Saavedra, A.; Sarti, G. C. (1997), ''Vacuum Membrane Distillation: Experiments and Modeling'', *AIChE J.*, Vol. 43, No. 2, pp. 398-408.

466009 Isolation of Biobased Chemicals By Combination of Reactive Distillation and Pervaporation; Process Modeling

Abstract:

Wastewater from pulping contains several acids, mainly acetic acid and formic acid. The ternary mixture acetic acid, formic acid and water forms a binary high boiling azeotrope between formic acid and water and a ternary saddle point azeotrope. The separation step of azeotropic mixtures is energy intensive, especially when water is present as low boiling component with high excess. By changing the substance properties via esterification, reaction products with lower boiling points and low enthalpy of evaporation compared to water as well as the acids are produced.

Reactive distillation is perfectly suited for the separation of acetic acid, formic acid and water by esterification with methanol or ethanol. Therefore low boiling esters are produced which are continuously removed via the distillate. Elevated temperature in the reboiler and simultaneous esterification of both acids enhances the reaction rates. High conversion of the acids is achieved. The distillate contains the esters and excess alcohol. This mixture can further be separated via pervaporation by removing the alcohol with hydrophilic membranes. Chemical conversion of the carboxylic acids simplifies the separation step and produces esters with a higher market value.

Minimization of energy consumption and maximum acid conversion of reactive distillation was optimized by simulation and validated experimentally. Diffusivity of membranes in pervaporation of binary and ternary mixtures was determined experimentally and was modeled with an activity based approach.

467093 Modelling and Optimization of Multistage Separation Process at Oil Offshore Platform

Abstract:

In the era of low oil price, major oil companies strive hard to improve accuracy of the offshore process design and operation efficiency of the platform. For this, this study focuses on the development of optimized offshore topside model, satisfying product specification, maximizing oil production and consequently maximizing profit. First, modeling of topside process that includes three phase separators, gas compressing system and recycle scheme was conducted, and optimal process configuration was selected based on evaluation of multiple case studies. Then, optimization of operating condition (e.g. gas to oil ratio, operating pressure, etc.) was performed by using Mixed Integer Non-Linear Programming (MINLP). Lastly, economic feasibility study was conducted for the design of profitable and eco-friendly offshore platform.

469782 Dynamic Intensification: Periodic Operation of Intensified Process Units

Abstract:

Dynamic Intensification: Periodic Operation of Intensified Process Units

Richard Pattison, Austin Dulaney, Michael Baldea

McKetta Department of Chemical Engineering

The University of Texas at Austin, 1 University Station C0400, Austin, TX 78712

email: mbaldea@che.utexas.edu

One of the core tenets of process intensification is to "do more with less." This includes carrying out more operations in fewer units (integration) and replacing conventional units with equal throughput units with smaller physical dimensions (miniaturization).

In a different vein, periodic operation of process units, particularly chemical reactors, was shown to increase throughput for many reaction schemes in comparison to steady state operation of a plant of the same capacity [1]. However, the potential benefits of periodic operation are often limited by material and energy transport constraints. Intensified units with small physical dimensions that combine both reaction and separation can eliminate some of these physical constraints by increasing heat and mass transfer rates through high surface area to volume ratios.

In this paper, we focus on the broad class of intensified processes that feature reaction and separation, such as reactive distillation columns and membrane reactors. We develop a prototype system representation under the assumption on unconstrained inlet and outlet material fluxes [2] (i.e., the reacting system is not limited by transport phenomena, which provides a design target for intensified unit operations). On this basis, we utilize the Pi criterion to define conditions under which imposing an oscillatory behavior (e.g., by periodically varying the manipulated variables) results in superior productivity compared to steady state operation. Our results are based on analytical solutions for several common reaction schemes, including reversible, consecutive and higher-order reactions.

We illustrate our theoretical framework with several detailed case studies. Moreover, we use numerical analysis to tackle the more complex cases whereby the material fluxes to the intensified unit are constrained. Our simulation results show that intensified reaction-separation systems can benefit from periodic operation for most reaction schemes, proving that dynamic intensification is a highly promising avenue for advancing the state of the art in the process intensification realm.

We will present the framework for a general reaction scheme and provide analytical conclusions for specific cases, then present results of detailed and optimized case studies, where we no longer assume that the material flux is unconstrained.

[1] Sterman, L.E.; Ydstie, B.E. Periodic forcing of the CSTR: An Application of the generalized Π criterion. *AIChE J.*, **1991**, 37, 986-996.

[2] Peschel, A.; Freund, H.; Sundmacher, K. Methodology for the design of optimal chemical reactors based on the concept of elementary process functions. *Ind. Eng. Chem. Res.*, **2010**, *49*, 10535-10548.

460244 Process Intensification and Distillation: How to Control a Fixed Design Dividing Wall Distillation Column

Abstract:

An example of one of the emerging technologies in the field of process intensification is dividing wall distillation columns or DWCs. As part of an ongoing experimental study of dividing wall columns at The University of Texas at Austin, the design and selection of control strategies for a fixed design DWC will be presented. A dynamic process simulator was used to test various feed compositions on a dividing wall column to understand the operational flexibility of a fixed design DWC. A fourth trace component was added to the feed and the selection of and changes in manipulated variables necessary to move the trace component between product tanks was investigated. Conventional distillation temperature control techniques were tested on the dividing wall column as well as more advanced control strategies, such as model predictive control, thought to be better suited to capture the inherent interactions of the system caused by the intensified nature of the process. The ability of these control systems to handle setpoint changes and feed disturbances was compared. To verify the results from the simulation, the control systems were implemented on a pilot scale dividing wall distillation column from which dynamic responses were obtained.

460843 Techno-Economic Analysis of a Single-Pass Methanol Synthesis Process

Abstract:

< style="padding-left: 40pt; text-align: center;">Techno-Economic Analysis of a Proposed Hybrid Methanol Synthesis Process

K.M. Bentaher, R.J. Machiela, T.N. Rogers(*)

Michigan Technological University, Department of Chemical Engineering 1400 Townsend Drive, Houghton, MI 49931-1295

(*)Corresponding Author

< style="padding-left: 5pt; text-align: left;">Abstract

Recent advances in drilling technologies, such as horizontal drilling and hydraulic fracturing, have led to shale gas revolution and considerably boosted U.S. natural gas production. Between 2005 and 2011, U.S. dry gas production increased by about 27 percent. This significant increase in the supply of natural gas has driven down its spot market price by more than 50 percent in the past 7 years. Since methanol is a key domestic transportation fuel, gasoline additive, and intermediate for many other important industrial chemicals, methanol production methods that employ a natural gas feedstock are re-evaluated in this paper to take advantage of the sharp drop in natural gas prices.

In this study, the starting point is the conventional methanol synthesis process which has a recycled syngas stream merging with the methanol synthesis reactor's feed stream. This recycle loop requires a compressor that represents a significant operating expense. To eliminate the syngas recycle loop, a new

hybrid process is proposed (Figure 1) that consists of a conventional methanol synthesis reactor (R-1) followed by a reactor (R-2) that condenses methanol vapor as it forms to achieve nearly complete one-pass conversion of syngas to methanol. By condensing methanol vapor on a cooling surface that spirals down the reactor's interior, the equilibrium limitation on methanol conversion is removed. The small amount of syngas that is not converted, rather than being recycled, is used as fuel in the reformer. The upstream conventional methanol synthesis reactor R-1 is included to decrease the R-2 reactor's feed rate, thereby reducing its size and cutting its high capital cost. The novelty of the new hybrid process is in the arrangement of conventional, proven equipment and technologies.

After-tax discounted cash flow analysis is used in this study to incrementally compare the new hybrid process design to two alternative approaches: conventional methanol synthesis with recycle and a process employing a single pass condensing reactor only. Cash flow analysis is also used to establish the optimum ratio of the sizes of reactors R-1 and R-2. Several standard metrics for profitability, such as NPV, ROI, IRR, and payback period, are reported for each candidate process over a 10-year analysis window.

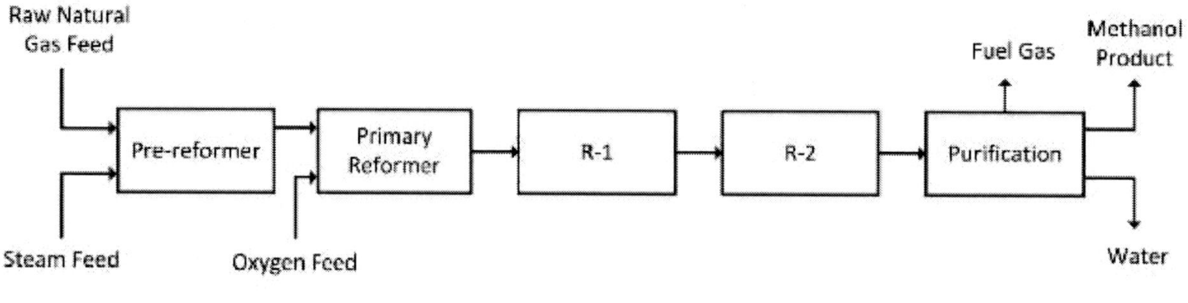

Figure 1. Proposed Hybrid Process for Converting Natural Gas to Methanol

461895 Comparison of Light Gas Separation By Membranes and Conventional Technology: A Case Study

Abstract:

In this presentation, an energy-intensive separation of light gases is selected as a case study, to compare the advantages and weakness of membrane-based separation with conventional technology. The study will determine the target membrane property (e.g. permeance, selectivity) for a variety of process configurations, and provide guidelines for membrane material development.

Continuous Vacuum Filtration Technologies for Sulfuric and Hydrochloric Acid Slurries

Garrett Bergquist, BHS-Sonthofen Inc., Charlotte, NC

Introduction

In many chemical manufacturing processes there exist slurry streams where the separation of solids from liquids is required. The range of equipment available for these streams is vast and the decision between equipment choices is not always easy. For clarification of low solids streams there exists technology such as candle filters, plate filters, and cartridge filters. These are typically a batch operation but can be made continuous by using multiple units in parallel. While one unit is actively filtering, the second unit is on side time operation such as filling, draining, or discharge. For higher solids applications continuous filters can be employed, including centrifuges, rotary vacuum filters, and horizontal belt filters.

Regardless of the specified filter type, when designing the filter, consideration must be made in the selection of appropriate materials. Materials of construction should be selected to withstand the temperature and chemical corrosion properties of the surrounding environment and liquids that come in contact with the filter.

This article discusses the use of alternate materials available in the construction of vacuum belt filters to address chemical compatibility concerns in applications using sulfuric acid and hydrochloric acid as the liquid constituent in a slurry. This article continues with a case study where the challenge is to process 2,200 kg/h dry solids that have been precipitated from a sulfuric acid solution.

Sulfuric Acid

Sulfuric acid is commonly and widely used in the production of fertilizer. Other manufacturing processes employ sulfuric acid to produce automotive batteries, pigments, iron, paper pulp, etc. Sulfuric acid can be a tricky liquid when it comes to designing robust systems for chemical compatibility. At high concentrations, sulfuric acid acts as a strong oxidizing agent. At low concentrations sulfuric acid acts as a reducing agent. For that reason, a material that is suitable for sulfuric acid at a high concentration may not be suitable for sulfuric acid at a low concentration and vice versa. Even at temperatures below 140°F (60°C), sulfuric acid at concentrations between 30wt% and 75wt% will corrode 316 stainless steel at rates greater than 200 mils per year (Figure 1).

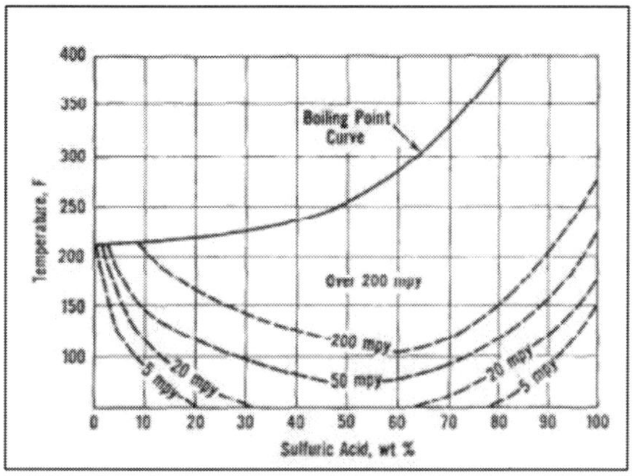

Figure 1. Corrosion of 316 Stainless Steel by sulfuric acid as a function of temperature. (Fontana, 1952)

When it comes to safety, sulfuric acid is an irritant and is considered very hazardous. In cases of skin contact and eye contact, burns may occur. Inhalation can result in severe irritation of the respiratory tract. Engineering controls are recommended to prevent physical interaction and to keep airborne concentrations of vapors below 1 mg/m^3 (NIOSH, 1997).

Hydrochloric Acid

Hydrochloric acid, also known as muriatic acid, is widely used in steel pickling, food production, calcium chloride production, gold mining, catalyst regeneration, dyes and pigment production, etc. Like sulfuric acid, hydrochloric acid is extremely corrosive to metals, including carbon steel, stainless steel, nickel and aluminum. In processes using hydrochloric acid, piping is often lined in thermoplastic and storage tanks rubber lined or made from FRP.

Hydrochloric acid is a highly hazardous substance and corrosive to mucus membranes. Exposure, even to vapors, can cause severe burns to exposed skin. The American Conference of Governmental Industrial Hygienists (ACGIH) recommends that the threshold limit value (TLV) for hydrochloric acid vapors to be 2 parts per million (ACGIH, 2003).

Vacuum Filtration Technology

In all vacuum filtration technologies, the concept and mechanism of filtration is the same. Vacuum is pulled through a filter cloth backed by a porous supporting structure while slurry is fed to the cloth. The liquid filtrate is pulled through the cloth by the vacuum pressure differential while the solids are retained on the surface of the cloth. The gas and liquid that pass through the cloth are then piped together to a gas/liquid separator to allow the liquid to fall via gravity which is then pumped from the bottom of the receiver. The separated gas is piped from the top of the separator to the vacuum pump and subsequently exhausted (Figure 2). The configuration in which this process is accomplished is what sets apart the different vacuum filtration technologies.

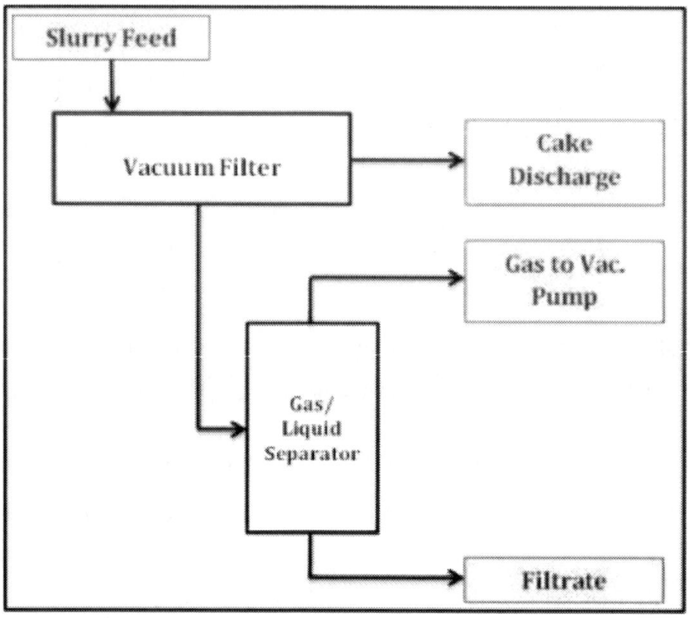

Figure 2. Process flow diagram for basic vacuum filtration

Rotary Vacuum Drum Filter

In a rotary vacuum drum filter, filter cloth surrounds a cylindrical segmented drum. The drum itself is constructed from an alloy material and acts as the porous supporting structure for the filter cloth. The bottom part of the drum is submerged in an open slurry trough. With vacuum being pulled through the cloth, filtration begins to occur as the drum rotates into the slurry trough. Cake heights are determined by the residence time spent submerged in slurry. As the drum rotates the cake can undergo washing via spray nozzles and dewatering by pulling residual moisture from the cake via gas flow from the vacuum pressure differential. The cake is then discharged from the media by a scraper knife or discharge roller.

Benefits in using a vacuum drum filters include having a large amount of filter area in a small footprint relative to horizontal belt filters. Additionally with the discharge design, a cake heel can be left on the drum which allows for slight depth filtration at the beginning of the following filtration cycle. When it comes to operation using hazardous chemicals, specifically sulfuric acid and hydrochloric acid, the vacuum drum filter has some major shortcomings. For one, the slurry trough and acid rich filter cake are open to the surrounding environment. If the acid is at all heated above ambient temperature, then a significant quantity of acid vapor will fill the surrounding environment. Given the height of the filter, providing a ventilated enclosure to keep acid fumes below the respective exposure limits is both difficult and expensive.

A second drawback comes when considering compatible materials for the vacuum belt filter piping and drum. Stainless steel alloys such as 304 and 316 are not suitable for use in sulfuric and hydrochloric acid solutions. As previously stated, with relatively low temperature a broad range of sulfuric acid concentrations can corrode stainless steel over 200 mils per year. This corrosion consideration leads to the material of construction specification of higher grade alloys such as Alloy 20 or Hastelloy. By using more exotic alloys the cost of the filter is increased significantly. When comparing the cost of Alloy 20 and Hastelloy C-276, the relative cost is over 3 and 7 times that of 316 Stainless Steel respectively (Grocki, 2012).

Continuous Horizontal Rubber Belt Filter

In the continuous horizontal rubber belt filter, an endless filter cloth is backed by an endless, perforated rubber drainage belt which in turn is supported by a deck attached to the filter frame. The supporting deck is made of rollers or the rubber belt is floated by blowing air or slide water to reduce friction and wear as the rubber belt travels across it. The surface of the rubber belt is grooved to allow filtrate flow towards the belt perforations which are sealed to a vacuum box underneath the rubber belt. Motion of both the rubber belt and filter cloth are continuous at a constant speed.

The horizontal rubber belt filter is well utilized in fast settling slurries as the solids are fed onto the top of the belt vs the bottom feed mechanism of the vacuum drum filter. Additionally if intensive cake washing or longer drying times is required, the length of the belt filter can be designed to accommodate these process requirements.

When considering the use of the horizontal rubber belt filter for slurries with sulfuric and hydrochloric acid, some of the concerns regarding environmental exposure and material of construction can be addressed. A fume hood over the filter can be constructed more easily given the horizontal surface. However since the rubber belt must periodically be replaced via the side of the machine, the hood can only be suspended over the filter and not attached to the frame. Curtains can be hung from the edges of the fume hood to cover the remaining space but do not eliminate the risk of someone reaching in to the filter or overflow at the slurry feed zone in an upset condition. Additionally, to ensure material compatibility, different grades of rubber can be selected for the drainage belt material. However for sulfuric and hydrochloric acid, standard rubber materials such as natural rubber, SBR, and NBR are incompatible to various degrees. Like the vacuum drum filter, higher grades of materials must be considered for the rubber belt such as EPDM (Moss Rubber, 2005). Using these higher grade materials add significant cost to what is already most expensive piece of the rubber belt filter.

Indexing Belt Filter

In the indexing belt filter, an endless filter cloth is backed by a stationary perforated filter bed. During operation the belt moves only periodically, step-wise down the filter bed. When the belt is stationary, vacuum is applied through the filter media to allow filtration. When the belt is indexing the vacuum is vented to allow motion across the stationary bed.

Like the rubber belt filter, the indexing belt filter is well utilized in fast settling slurries. Intensive washing can be performed like the rubber belt filter. There is a slight advantage for the indexing belt filter when it comes to countercurrent wash schemes as the wash additions and filtrates are better separated with the indexing design. Also with the indexing filter since the belt is stationary for a period of time, a press can be utilized to squeeze the solids while simultaneously blowing air through the cake to remove all remaining free moisture. The drawback to the indexing design is that side time introduced during the atmospheric pressure belt movement which reduces the filtration capacity compared to a continuous rubber belt filter design. For that reason the indexing belt filter it is better suited for medium to slower filtering slurries compared to ideal continuous rubber belt filter applications.

When considering the use of this design for hazardous slurries, the indexing belt filter is a perfect example of flexibility in both environmental considerations and available materials of construction. Because the filter bed is stationary and the cloth can be changed from the discharge end of the filter, a full hood is able to be built over the entire surface of the filter without side gaps (Image 1). This is a significant benefit when designing the filter for sulfuric or hydrochloric acid slurries as it completely isolates any interaction with the product. Through the use of blower a slight negative pressure can be maintained over the filter at all times. Alternatively the indexing design can also provide for a full gas tight enclosure which is beneficial for filtration when explosion proofing must be considered. The gas tight enclosure is able to be nitrogen purged and maintained using a nitrogen recirculation skid.

When considering materials of construction, the indexing design is able to be built from a multitude of polymers which commonly include polyethylene, polypropylene, or PVDF. These materials are able to be used for 100% of the wetted surfaces of the filter. The chemical resistance against sulfuric and hydrochloric acid with this selection of polymers is excellent at all concentrations ("Chemical Compatibility Database", 2016). Additionally the material cost of these polymers compared to high grade alloys such as Alloy 20 and Hastelloy is much less. In fact, the use of polypropylene is even cheaper than construction using 316 stainless steel. The main limitation for using polymer constructed belt filters is temperature resistance. For a polypropylene belt filter, the maximum operating temperature is 175°F – 194°F (80°C-90°C) and 212°F – 230°F (100°C-110°C) for PVDF. However, for most vacuum filter applications this is not an issue as temperatures are often below the boiling point of water.

Image 1. BHS polypropylene indexing belt filter with oscillating slurry feed, fume hood and pressing/blowing device

Case Study – Sulfuric Acid Filtration

A company producing solid chemicals used in paper coatings was using a process in which product dissolved in sulfuric acid (H_2SO_4) is precipitated using a neutralization step via NaOH. This company was looking to move away from NaOH neutralization as the price of NaOH had increased significantly over the past year. An alternative production method was known in which the concentration of sulfuric acid was diluted with water to approx. 25wt%. The dilution then allowed the product to precipitate from solution. With this method any residual sulfuric acid must be removed from the solids as all downstream equipment was constructed of stainless steel. Additionally the precipitated solids are fragile so no intense agitating or centrifugation was possible. Benchtop scale testing using a 20cm^2 pocket leaf filter

(PLF) was performed to determine the appropriate technology, sizing, wash requirements and filter cloth to meet the application requirements.

PROCESS DATA

Slurry Throughput:	12,500 l/hr	(55 gpm)
Slurry Density:	1.3 g/mL	(30% H_2SO_4)
Suspended Solids:	13.5 wt%	
Dry Solids Throughput:	2,200 kg/h	
H_2SO_4 Concentration:	22-30 %	
Slurry Temperature:	max. 140°F	(60°C)

Through benchtop scale testing for the vacuum belt filter with the 20cm^2 PLF the following data was measured:

Filter Cloth:	80µm polypropylene
Test Filter Area (A_f):	0.002m^2 (20cm^2)
Slurry Feed:	425g
Filtration Time:	11 seconds
Mother Filtrate Quantity:	175 mL
Wash Volume:	235 mL
Wash Time:	55 seconds
Wash Filtrate:	238 mL
Drying time:	20 seconds
Heel Filtrate:	10mL
Cake Height:	50 mm
Wet Cake Weight:	185 g
Residual Cake Moisture:	65 wt%
Dry Cake Weight (m):	65 g

It was determined through preliminary calculation that a BF100 with 1.0 meter wide active belt width and an area of A_{zone} = 0.75m^2 per zone was appropriately sized. The cycle time and zone arrangement scale up for the BF100 at a filter cake height at 50mm was performed as follows:

Solids loading per stroke:

$$\frac{m}{A_F} * A_{Zone} = \frac{0.065kg}{0.002m^2} * 0.75\frac{m^2}{Stroke} = 24.38\frac{kg}{Stroke}$$

Number of strokes per hour:

$$N = \frac{2200kg/h}{24.38kg/Stroke} = 90Strokes/h$$

Duration of one cycle:

$$t_{Cycle} = \frac{3600\frac{s}{h}}{90\frac{Strokes}{h}} = 40\frac{\sec}{stroke}$$

With an indexing time of 5 sec with no applied vacuum, the active processing time per cycle is 35 sec. The belt filter zone arrangement based on an active time of 35 seconds is recommended as follows:

Table 1. Indexing Belt Filter BF100 Zone Arrangement

Process Steps	Test Times	Theoretical No. of Zones	Actual No. of Zones
Feed	-	-	1
Filtration	11 sec	0.31	1
Wash	55 sec	1.57	2
Drying	20 sec	0.57	1
Total	-	-	5 zones

With a total number of 5 zones required at $0.75m^2$ each, a belt filter BF100-038 with $3.75m^2$ in filter area was recommended to process 2,200 kg/h dry solids from the sulfuric acid solution (Image 2). A feed zone at the beginning of the filter with no vacuum was also recommended to allow for a more uniform solids distribution.

Image 2. Polypropylene belt filter with fume hood built for the filtration, washing, and drying of the sulfuric acid slurry.

Conclusion

When considering new technology or a new production processes it is always important to verify that the specified materials match the chemical and temperature resistance requirements for each stream. While the technology in the market place for vacuum filtration is often well known, when it comes to dealing with hazardous slurries such as those containing sulfuric or hydrochloric acid the options should be carefully considered. Production areas containing acids should be fully enclosed to prevent splashing or spills in an upset condition, and to capture any hazardous acid vapors. Special attention must be made to specify compatible materials as stainless and carbon steels are often incompatible with most sulfuric and hydrochloric acid concentrations. However specifying compatible specialty alloys such as Alloy 20 and Hastelloy for wetted parts becomes very expensive quickly. With traditional filters such as the rotary drum filter or the continuous rubber belt filter providing containment and identifying economic compatible materials can be a near impossible task. With the indexing vacuum belt filter, fully enclosed or even gas tight operation is easily achievable. More importantly, through the use of polymer wetted parts such as polypropylene or PVDF the cost of the filter is able to be kept low, even lower than 316 stainless steel construction. Polymer constructed belt filters in hazardous acid environments have a proven track record of safe and reliable operation with low costs of purchase.

References

1. ACGIH: Documentation of the Threshold Limit Values (TLVs) and Biological Exposure Indices (BEIs) - Hydrogen Chloride. 2003.
2. "Chemical Compatibility Database." Chemical Compatibility Data Base. Cole-Parmer, n.d. Web. 19 Sept. 2016.
3. Fontana, M.G., "Corrosion Series," Ind. Chem. Eng., *44* March 1952, P89A
4. Grocki, John. "Relative Cost 10-12." Pressure Vessel and Tank Fabrication Using Stainless Steels. Proc. of Steel World Expo, Houston. STI/SPFA, Oct. 2012. Web.
5. Moss Rubber and Equipment Corp. Chemical Resistance Chart. Benicia: Moss Rubber and Equipment, 2005. Print.
6. National Institute for Occupational Safety and Health. (1997). *NIOSH pocket guide to chemical hazards*. Cincinnati, Ohio: U.S. Dept. of Health and Human Services, Public Health Service, Centers for Disease Control and Prevention, National Institute for Occupational Safety and Health.

462318 Improved Olefin Polymerization Process with Reduced Energy Consumption

Abstract:

Improved Olefin Polymerization Process With Reduced Energy Consumption

Carrying out olefin polymerization reaction in solution has some distinct advantages such as a homogenous environment to facilitate consistent fast reaction; solvent acting as natural heat sink to absorb heat of polymerization, and fast product grade transition. On the other hand, the need to remove and recover large amount of solvent used in solution polymerization reaction, and the requirement to meet target solvent residues in the final product, greatly increases the capital expenditure and energy consumption of the designed asset.
The conventional method to remove solvent in a solution polyolefin process is by heating up the polymerization reactor effluent to desired temperature, then flashing off the solvent through vapor liquid equilibrium (VLE) using multiple devotalization stages at different operating pressures, which normally require deep vacuum in the final vapor liquid separator. The amount of solvent that needs to be processed by the devotalization and solvent recovery section of the plant dictates the maximum production throughput, and leads to higher energy cost than other alternative olefin polymerization process.

The approach of applying Lower Critical Solution Temperature (LCST) behavior of polymer solution2, and subsequently inducing a "front end" liquid-liquid separator of polymer solution followed by one or two finishing vapor liquid separator, has been explored in the recent years. Under carefully designed temperature, pressure and solvent environment, a homogenous polymer solution may be split into two liquid phases: a polymer rich phase and a solvent rich phase. This almost isothermal separation process1 has some unique characteristics that can be advantageous to designing an energy efficient solution polyolefin manufacturing process. However, achieving sufficient and reproducible liquid-liquid separation can be difficult if the solvent composition does not possess the needed high volatility, or the system temperature is not substantially higher than the Lower Critical End Point (LCEP) of the reactor effluent.

This presentation explored the key thermodynamic aspects of polymer solution Liquid-Liquid Equilibrium (LLE), and described a unique approach to achieve effective liquid-liquid separation. Unlike the conventional process of using solvent with a fixed boiling point, It applies a mixed solvent composition of one high boiling point solvent, and a low boiling point solvent. The low boiling solvent acts as the anti-solvent that significantly alters the LCST behavior of the polymer solution and reduces the LCEP of the system by 40 ~ 60 °C. This mixed solvent approach enables effective polymer-solvent separation inside the liquid-liquid separator without the need to put out energy to heat up the polymer solution stream exiting the reactor. In addition, the amount of low boiling solvent (anti-solvent) can be adjusted and used as a control knob to ensure a consistent liquid-liquid separation outcome (i.e., similar polymer concentration in the polymer rich phase). The amount of low boiling solvent can also be used to adjust for changes in compositions or reactor conditions (e.g. different product design or buildup of inerts).

This new liquid-liquid separation based Olefin Polymerization Process3 brings a 35% reduction of the total plant energy input as compared to conventional technology using vapor liquid separator alone, because heat is added only to the polymer-rich phase that is downstream of the Liquid-Liquid Separator (LLS). In addition, the cooling requirement for the solvent-rich phase downstream from the LLS is also lower as no heat was used to heat up the reactor effluent. Moreover, this new design also brings capital savings as it enables the use of lower cost adiabatic reactors operating at lower polymer concentration.

Reference:

1. TG. Gutowski, NP Suh, C Cangialose, and GM Berube, "A Low-Energy Solvent Separation Method," Polym Eng & Sci 23:230-237, 1983
2. Freeman, P.I. and J.S. Rowlinson, Lower critical points in polymer solutions. Polymer, 1960. 1: p. 20-26.
3. P. Jog, Midland, R. Swindoll, N. Mead, P. Jain, A. Wang, J. Guzman, United States Patent 9018328, Polymerization process for olefin-based polymers.

472392 On the Dynamics of Non-Coalescing Microscale Bubbles in Turbulent Flows for Intensification of Gas-Liquid Mass Transfer

Abstract:

In many gas–liquid contactors or reactors, mass transfer from the gas phase to the liquid phase is a major player in the reactor's overall production and/or economics. The mass transfer rate can be enhanced either by a) increasing interfacial area *a*, b) intensifying the mass transfer coefficient k_L, or c) increasing the solubility of gas via higher pressure or complexing reagents. All three of these beneficial phenomena may be leveraged simultaneously by use of micron-scale bubbles. Therefore, to intensify reactor mass transfer rate, we studied the behavior of microbubbles (< 200 micron) in a 100L bubble column and in a 500 mL bench top column. Specifically, the stability of microbubbles against coalescence was studied under additions of surfactant (sodium dodecyl sulfate (SDS) [$CH_3(CH_2)_{11}OSO^{-3}$ Na^+]) and salt (KCl), both of which are known to decrease bubble coalescence (Parmar & Majumder 2014). According to previous literature, adding salt to the liquid phase in the presence of SDS has two different effects on the bubble surface tension and the hydrostatic repulsion between bubbles (Prosser & Franses, 2000): a) Na^+ ions allow closer packing of surfactant molecules at the gas-liquid interface, which diminishes surface tension because surface tension is inversely proportional to the density of surfactant molecules at the surface, and b) decreases the repulsion between bubbles because the electrostatic Debye length decreases. Depending on the salt concentration, the stability can be changed from stable to unstable. Low concentrations of SDS (0-35) ppm w/w and KCl (0-75) ppm w/w were added to the solution. We observed that the presence of SDS up to 35 ppm w/w and KCl up to 75 ppm w/w did not alter surface tension by more than a few percent. However large change was observed in mean bubble size, k_La, and bubble coalescence rate. It was hypothesized that the stability of microbubbles was not from the slight decrease in the surface tension but from an increase in zeta potential. Zeta potential (ζ) was measured by high-speed video of the time-dependence of bubble motion inside a 50 cm tall bubble column with an oscillating electric field applied across the bubble column, similar to a method used previously by Takahashi (2005). Bubble size and coalescence rate were also measured. Results showed a significant increase in ζ-potential of bubbles as SDS and KCl concentrations were increased. An energy balance confirms ζ-potential to be the leading factor for decreased coalescence rates and decreased bubble size. In a test reactor, k_La and *a*values increased both by 50% as SDS and salt concentrations was increased, likely due to an increase of interfacial area.

References:

Parmar, R. & Majumder, S. K. 2014 Hydrodynamics of microbubble suspension flow in pipes. Ind. Eng. Chem. Res. 53, 9, 3689–3701.

Prosser, A.J. & Franses, E.I., 2000 Adsorption and surface tension of ionic surfactants at the air–water interface: review and evaluation of equilibrium models. Colloids and Surfaces A: Physicochemical and Engineering Aspects. 178, 1–40.

Takahashi, M. 2005 ζ Potential of Microbubbles in Aqueous Solutions: Electrical Properties of the Gas-Water Interface. J. Phys. Chem. B. 109, 21858-21864.

468032 Mathematical Modelling of Liquid â€"Liquid Extraction in the Slug Flow Regime in a Microchannel

Abstract:

Slug flow is a commonly observed flow regime in the microchannel. Mixing in the slug flow regime is enhanced by internal circulations induced by shearing due to wall. This helps improve mass transfer in this flow regime.

In this work, we exploit the low Reynold's number characteristic of the flow and seek a numerical solution to understand the structure of the vortex patterns formed in the two phases in the slug flow regime. Further, we study liquid-liquid extraction in the system and determine the improvement in mass transfer performance. A Lagrangian approach where we analyze the behaviour of a unit cell in the channel in a moving reference frame is adopted for the analysis. The system is analysed for two cases when there is (i) no film surrounding the slug (ii) a thin film surrounding the slug. The 2D governing equations for fluid flow are solved using two approaches: a) a stream function formulation based on finite differences, b) primitive variable formulation with the Chebyshev collocation method. Results obtained from the two methods are validated with each other. Then, the species balance equation is solved numerically using a finite volume code. The extraction performance is analysed in terms of extraction efficiency and mass transfer coefficient. It is believed that the enhancement in mass transfer is due to the mixing induced by the strong internal circulation set up due to the channel wall. The model is validated with published experimental results.

464382 Mass Transfer Enhancement as a Function ofÂ the Oscillatory Baffled Reactor Design

Abstract:

The oscillatory baffled reactor (OBR) is an intensified design of plug flow reactor (PFR). It is able to achieve a high degree of plug flow at very low flow rate due to the interaction between the oscillatory flow and baffles which improves mixing inside the OBR. It was found that superimposing fluid oscillation in a baffled column enhanced the mass transfer significantly comparing with many devices, for instance, bubble column, internal and external loop column, and stirred tank. This is due to the highly turbulent flow pattern that generated in the OBR as a result of the oscillatory motion which leading to frequent bubble breakage, a subsequent reduction in the Sauter mean diameter, an increase in bubbles retention, when trapped in recirculating vortices, and the gas-liquid contacting area

Recently, oscillatory baffled column (OBC), has used for the same purpose, mass transfer, and presented a significant change in which has been increased up to sixfold comparing with a bubble column. This was due to the combanation effect of baffles and oscillatory motion that increase the radial mixing (diffusion). Also gas-holdup affected by the oscillation frequency and amplitude.

In this study, the volumetric mass transfer coefficient ($k_L a$) in air-water system has been calculated and compared using three designs of OBRs, smooth periodic constrictions (SPC) design, helical baffled design, and single orifice design. This work was conducted over a wide range of oscillation condition and aeration rate, volumetric air flow rate. The SPC presented higher values of $k_L a$ than the single orifice baffled reactor and the helical baffled reactor. This indicates that SPC with oscillatory motion provide bigger interfacial contact area between air and water. In addition, the visualisation of bubbles movements inside the three designs of OBR has shown that the bubbles collision and breakage in SPC was more severe. Therefore the bubble size in SPC was smaller than those that achieved in the single

orifice OBR then in the helical OBR.

455522 Isolation of Carboxylic Acids from Dilute Aqueous Solutions – Intensifying Approach By Conversion in Emulsified Reactive Solvents

Abstract:

The focus on the production of biobased bulk chemicals as well as the progress in the biorefinery require innovative downstream processing of the product effluents. Examples are provided by different cellulose pulping processes with varying yields of carboxylic acids (e.g. acetic acid or formic acid). The effluents from such processes are usually characterized by low concentration of the constituents in aqueous solution. Conventional separation techniques quickly reach economic limits. The requirements for downstream processes capable of dealing with such feed specification, must suffice effective separation to ensure economic utilization of the valuable constituents.

The intensification of conventional separation techniques by combining with chemical reaction provides high potential in overcoming these obstacles. Two approaches, reactive distillation for the isolation of carboxylic acids and chemical conversion of the target constituents in both phases of a liquid-liquid extraction process, may offer a feasible option. Reactive conversion of choice for isolation of carboxylic acids is the esterification with an aliphatic alcohol, whereby the market value of the carboxylic acids can be significantly increased and the significant change of physical properties favorably influences the separation process.

As the partition behavior of the reactants is supportive when using higher aliphatic alcohols (e.g. *n* butanol or *1*-octanol), the use of these alcohols, either dissolved in corresponding esters or hydrocarbon based solvents has been investigated. Catalysis and elevated temperature are provided to increase the reaction rate of the esterification reaction. *4* dodecylbenzenesulfonic acid (4DBSA), a strong anionic surfactant and highly active homogeneous catalyst, has been used in the experimental validation in order to highly increase the contact between the reactants and the catalyst by emulsification. After having finished the reaction the emulsion is separated from the aqueous phase by settling and split by coalescence. A continuous process has been designed by altering conversion, phase separation, emulsion splitting and isolation of the product. Exemplarily 80% separation efficiency was obtained for acetic acid separation from aqueous feed by esterification with octanol in the emulsified solvent phase in single stage isolation.

474643 High-Temperature Ethane Dehydrogenation in Microporous Zeolite Membrane Reactor

Abstract:

Ethene is the largest base chemical for the chemical industry and produced either by cracking or by dehydrogenation of light alkanes. The increasing demand for ethene has stimulated substantial research into the development of new processes to reduce energy consumption. Catalytic ethane dehydrogenation using a membrane reactor is an attractive solution because the cracking equilibrium can be shifted in favor of ethylene by selective removal of hydrogen.

In this study, we report the intensification of ethane dehydrogenation reaction in packed-bed membrane reactors operating with a Pt/Al_2O_3 catalyst. The effects of tubular MFI zeolite membranes on ethane conversion and ethene selectivity are investigated. The rate of H_2 membrane permeation significantly enhances the ethane conversion. MFI membrane reactors allow the equilibrium limit of ethane conversion to be surpassed at high temperatures. It is demonstrated that medium-pore MFI membranes with moderate H_2/C_2H_6 selectivity can be effective in enhancing ethane conversion at high operation temperature by timely removal of H_2 through the membranes.

464166 Analysis of a Cyclic Redox Process for Oxidative Dehydrogenation of Ethane

Abstract:

Ethylene is commercially produced from ethane via high-temperature pyrolysis in the presence of diluting steam a.k.a. steam cracking. The process requires a high temperature (>1100 K) and a large steam generation load, making it highly energy intensive (17-21 GJ/tonne ethylene). The process also emits significant amount of NO_x and CO_2 and requires periodic shutdowns for coke burnout. Moreover, single-pass ethylene yield is limited by thermodynamic equilibrium. Oxidative dehydrogenation (ODH) of ethane, where ethane is partially oxidized to ethylene and water, represents an alternative route. The partial burning of the product hydrogen provides the required heat while shifting the equilibrium towards the product side. As a result, higher single-pass ethylene yield can potentially be achieved with reduced energy consumptions. The requirements for highly-selective catalyst, co-feed of combustible oxygen and ethane mixture, as well as cost associated with oxygen separation represents key challenges for such a process. A chemical looping based ODH (CL-ODH) scheme addresses the abovementioned challenges. In this process, the required oxygen is supplied by lattice oxygen from a redox catalyst composed of mixed metal oxides. The reduced catalyst is oxidized with air in a separate reactor, where it is regenerated and then recirculated to the ODH reactor. This allows built-in air separation with minimal parasitic energy loss and better temperature control. Overall, CL-ODH transforms the upstream of a traditional steam cracking process and has the potential to enhance ethylene production with significant energy savings.

Mn-based oxides have been shown to be effective as redox catalysts for the ODH of ethane. Selective combustion of hydrogen to water, over the redox catalyst, shifts the equilibrium towards ethylene and also provides the heat required for the endothermic dehydrogenation reaction. In this work, the ODH of ethane is modeled using ASPEN Plus® and is compared with a conventional stream cracking process. Results show that CL-ODH with 85% ethane conversion provides over 70% reduction in the overall energy demand with >75% reduction in the overall CO_2 emissions. The exothermic nature of the regenerator and elimination of the steam requirement lead to major reductions in the upstream energy consumption. Combustion of hydrogen further reduces the downstream separation loads by >20%, with >25% drop in the compression work. For every tonne of ethylene produced, steam cracking requires an external fuel input equivalent to 1.5 GJ_{Th}, whereas the CL-ODH results in surplus fuel energy in excess of 6 GJ_{Th}. Increase in single-pass ethane conversion further increases the energy savings. Preliminary results from CHEMKIN-PRO® are in close agreement with literature values in terms of product compositions for ethane cracking. Addition of surface kinetics of the redox catalyst towards hydrogen combustion provides further understanding of the dependence of product output on the process conditions.

474954 Novel Method to Isolate Polymeric Slurry Based Systems in a Continuous Process

Abstract:

During polymerization reactions, if the final polymer is not soluble in the solvent, a slurry or precipitate is formed as the molecular weight builds up. Typically, after the reaction is completed, a series of unit operation such as centrifugation followed by drying is used to isolate dry polymer. Slurry processes have inherent challenges with maintaining stable flow control for continuous operation; powder carryover with vapor; powder handling and other risks associated with potentially explosive dusts.

To describe the new process, we have used the example of a polymer that forms a slurry in the solvent most suitable to carry out its reaction. It was identified that above a certain temperature higher than the boiling point of the solvent, the polymer was soluble in the solvent. In order to form a solution, the reaction mixture would need to be maintained at an elevated temperature and corresponding vapor pressure. By having a polymer solution instead of a slurry, it was possible to use standard thermomechanical separators such as an agitated thin film evaporator (ATFE) or devolatization extruders to isolate the polymer in the continuous process. It was observed that the polymer would crash out of the solution if the temperature or pressure dropped below a certain limit. With this in mind, an overview of the risk mitigation activities undertaken to test the feasibility of reaction and isolation of the polymer in an ATFE operating under pressure is provided. A similar approach could be taken for isolation of other polymers, which are not soluble in a solvent below its boiling point, but are soluble above the boiling point and can be achieved under elevated pressure conditions.

455485 Scale-up of Oscillatory Baffled Reactors (OBRs)

Abstract:

Scale-up of oscillatory helically baffled reactors (OHBRs)

Safaa M.R. Ahmed, Anh N. Phan, Adam P. Harvey

School of Chemical Engineering and Advanced Materials, Newcastle University

Newcastle upon Tyne - NE1 7RU

Mobile:+44 7774101215

E-mail: s.m.r.ahmed@newcastle.ac.uk

Keywords: Oscillatory baffled reactor, scale-up, residence time distribution

Abstract

An Oscillatory Baffled Reactor (OBR) is an intensified design of continuous plug flow reactor (PFR) in which plug flow behaviour can be achieved at very low net flows (laminar flow regime). OBRs consist of tubes with periodically spaced baffles of various designs (orifice, helical, integral etc. baffles). There is a net flow through the reactor, and a superimposed oscillatory flow. The oscillatory flow interacts with the baffles to produce flow structures (usually vortices) that provide mixing. The mixing in the OBR is therefore independent of the net flow. As a result, the OBR's niche application is to operate "long" reactions in continuous mode. This is usually impractical in conventional tubular reactors.

Scale-up in conventional reactors, i.e. stirred tank reactors, is unpredictable due to the non-uniform mixing at large scale, leading to large variations in concentration, temperatures etc. This means that optimum conditions obtained from laboratory scales cannot be directly used at large scales, therefore process development/ product-to-market time would increase. However, scale-up of OBRs should be more predictable, as the flow structures can be reproduced across length scales.

Recent studies on oscillatory helically baffled reactors (OHBRs) at small scales (millilitre volume) found that the helical baffled design could provide high degree of plug flow across a wide operating window due to the combined effect of vortex formation and swirl flow. It was found that the behaviour of residence time distribution remained the same at all tested scales (10mm diameter to 25mm diameter, corresponding to volume 0.078L to volume 0.834L) when the similarity in geometric and dynamic parameters was maintained. The degree of plug flow was quantified in terms of number of tanks-in-series (N). At a fixed geometry, a scale-up correlation was established and validated over a range of operating conditions such as oscillatory conditions (Strouhal number, *St*, and oscillatory Reynolds number, Re_o) and the velocity ratio of oscillatory flow and net flow (ψ).

$$\frac{N}{N_o} = 40\, Str^{-0.3}\, Re_n^{\;-1}\, \psi e^{-0.1\psi} + 1$$

Where, $N_o = 0.25\, Re_n$

456248 Flow Microreactor Synthesis of ZIF-8 Particles with Controlled Size, Shape, and Adsorption Properties

Abstract:

Significant progress has been made in research and development of metal-organic frameworks (MOFs), which consist of metal ions and organic ligands. Their chief features include large surface area, highly ordered porous network, and structural flexibility. The flexibility leads to a structural transition of the host framework from a non-porous to a porous phase that causes a step-wise increase in guest molecules at a threshold pressure, which is called â€œgateâ€ adsorption. Because of this unique adsorption property, soft MOFs are promising for various applications not only to gas storage and separation, but also to molecular sensors. To exploit full potential of soft MOFs, their particle size and shape must be monodispersed because their unique adsorption property can depend on their size and shape. In the present study, we synthesized zeolitic imiazolate framework-8 (ZIF-8), which is a subclass of soft MOFs and is composed of Zn ions and 2-methylimidazole linkers. We mixed aqueous $Zn(NO_3)_2$ and 2-methylimidazole solutions in a central-collision type microreactor, which utilizes shear forces at the collision of two inlet fluids to realize rapid and homogeneous mixing. The mixed solution from the outlet was kept in a glass vessel for an hour. The nucleation starts upon mixing the solutions in the microreactor, followed by the particle growth in the glass vessel. Typical particle size is around 300 nm, and resultant particle size and shape distributions synthesized with the microreactor were more uniform than the case of a batch reactor. This would be due to a high mixing performance of the microreactor that can divide the nucleation process and the particle growth process. Moreover, we synthesized ZIF-8 particles for varied ratio of 2-MeIM to Zn^{2+} (R), concentration of Zn^{2+} ($[Zn^{2+}]$) and temperature. The average size of ZIF-8 decreases with the R, and a higher $[Zn^{2+}]$ results in smaller size at a fixed R, which is because of a higher degree of supersaturation at higher R and $[Zn^{2+}]$ conditions. Prepared ZIF-8 particles have BET surface area of 1500 m^2/g and exhibit the gate adsorption behaviour with different transition pressures depending on the particle size and shape. We also confirmed the microreactor synthesis is applicable to a different type of soft MOFs (ELM-12). In this manner, our flow process using the microreactor is demonstrated to be a simple but versatile technique for the synthesis of monodispersed MOF particles.

457239 Intensification of Batch-Distillation Processes for Production of Alcoholic Beverages

Abstract:

During production of alcoholic beverages, especially the alcoholic fermentation of pomaceous fruit, formation of methanol is an issue that has to be considered. Since the middle lamella which binds adjacent fruit cells, consists of methylated polygalacturonic acid, i.e. pectin, the formation of methanol is unavoidable during natural fruit specific enzymatic digestion. Furthermore pectin methyl esterase is commonly added to the fruit mash in order to gain an enhanced yield of ethanol. While methanol in principle is not toxic to human beings, the effect of metabolites formaldehyde and formic acid can cause severe injuries to health, e.g. blindness or acidosis and can even cause death. Thus the separation of methanol from distilled alcoholic beverages has to be explicitly taken into account.

Against the widely shared notion it is not possible to separate methanol by distillation using common batch distillation equipment, no matter if multi-stage columns or single-stage pot stills are used. Experimental investigations as well as dynamic simulations of single stage batch distillation with Aspen Hysys showed matching results. In fact methanol is distributed between foreshots, heart cut and tailings. Further the ratio of methanol to ethanol is increasing with cumulative distillation time. In addition high regulatory limits in the European Union for methanol concentrations in pomaceous fruit spirits of 1350 mg/100 ml pure ethanol compared to other spirits, e.g. wine brandies are often exceeded even though the high values are not set due to harmlessness to health but insufficient separation performance of standard batch distillation equipment. Hence the production of distilled pomaceous fruit spirits with less methanol content leads to the necessity of intensifying distillation processes with focus on new technical approaches.

The main goal of the current research is to improve the separation of methanol in the foreshots, using an extension of common lyne arms. For this purpose a pot still is equipped with an optional bypass which provides the possibility to route the vapor via an additional randomly packed lyne arm at the beginning of distillation to increase the separation efficiency. Further the bypass lyne arm enables enhancement of internal reflux, realized with a coaxial heat exchanger. In this manner methanol accumulates in the foreshots and may be separated from the heart cut sufficiently.

453255 Droplet Formation of H2SO4/Alkane System in a T-Junction Microchannel: Gravity Effect

Abstract:

A special system of concentrated sulfuric acid (H2SO4) and n-hexane was used to study the droplet formation in a glass T-junction microchannel with H2SO4 as the continuous phase. The effects of capillary number, flow ratio, and viscosity ratio on the droplet formation were investigated. The effect of gravity was explored by changing the flow direction in the microchannel. Results showed that the formation of transition flow pattern from squeezing to dripping is much easier for this special system compared with common aqueous/organic systems. This phenomenon is due to the considerably higher viscosity of H2SO4 than that of common aqueous phase and the higher density difference of the system compared with those of common systems. In addition to capillary number and flow ratio, gravity evidently affects the formation of droplets and flow patterns. The droplet size is smaller than that during the horizontal flow when the flow direction is consistent with gravity. By contrast, flow direction contrary to that of gravity results in larger droplet size than that at horizontal flow. This phenomenon provides guidance on the operation of these special systems in microchannels. Finally, mathematical models of droplet size at different flow patterns have been established, and these models can predict droplet size very well. This study could be helpful to extend the application of microreactors to new working systems.

AUTHOR INDEX